The Measurement Techniques of Space Geodesy

Optical technologies

Online at: https://doi.org/10.1088/978-0-7503-2032-0

The Measurement Techniques of Space Geodesy

Optical technologies

Karl Ulrich Schreiber

Research Unit Satellite Geodesy, Technical University of Munich, Munich, Germany

IOP Publishing, Bristol, UK

ISBN 978-0-7503-2032-0 (ebook)
ISBN 978-0-7503-2030-6 (print)
ISBN 978-0-7503-2033-7 (myPrint)
ISBN 978-0-7503-2031-3 (mobi)

DOI 10.1088/978-0-7503-2032-0

Version: 20251201

IOP ebooks

British Library Cataloguing-in-Publication Data: A catalogue record for this book is available from the British Library.

Published by IOP Publishing, wholly owned by The Institute of Physics, London

IOP Publishing, No.2 The Distillery, Glassfields, Avon Street, Bristol, BS2 0GR, UK

US Office: IOP Publishing, Inc., 190 North Independence Mall West, Suite 601, Philadelphia, PA 19106, USA

This book is dedicated to the pioneers of space geodesy who provided the foundation for a global geodetic frame of reference.

Contents

Preface

Assume you live near the sea and you go to the beach once every day. As the time goes on, you will notice that the sea level changes and a number of questions arise: 'Is the sea level going up?' or 'Is the landmass subsiding?' and 'Where is zero for height?' Simple as these questions may sound, there is no easy answer to them. Of course, we quickly realize, 'First of all, we require a reference frame against which we can monitor any changes.' But how do we set this up? Over many centuries artificial and natural landmarks served for this purpose. But these reference frames were not global and only suitable for localized communities. Owing to the limitations of the applied measurement techniques and the political circumstances, they degraded quickly as the distances got larger. Finally, these reference frames ended at the sea shore. Without any rigid landmarks in sight, there was no way to extend the mapping. When we look back over history, the available reference frames nevertheless suited the respective societies sufficiently well. Before industrialization, life was rather confined to local areas and those who traveled far followed the main trade routes. There was no need for elaborate navigation. Other, more serious problems dominated the life of the travelers in those days. Things began to change, when industries and trade reached higher volumes and extended to places much further apart.

With the advent of the satellite age, the field of space geodesy eventually emerged. Air traffic began to populate the sky and navigation was in high demand. Surveys did not only place cities, roads and natural resources on a map, they also allowed to see things changing. Now it was possible to develop a far-reaching survey grid for the first time that covered the entire globe. Satellites that more and more resembled the shape of a mathematical point mass, became external survey markers, which were able to look behind the horizon. However, this is about where the similarities end. This new type of survey marker no longer acquired a fixed position. In order to remain in the near earth orbit, it had to travel at the speed of several kilometers per second and the observation of such a reference marker became a highly dynamical task. The method of precise orbit determination had to be developed. In the end it turned out that these new fast-moving reference points in space could well take the role of the rigid geodetic markers of the past. The rapid development of the satellite technology was accompanied by other likewise rapid technological developments. The invention of pulse lasers generated another very significant opportunity. Now it was possible to utilize time-of- flight measurements, which proved to be much more accurate than the traditional angle measurements, in particular for the ground to space applications. This was the hour of birth of the laser ranging technique. Since the velocity of light is constant and at the same time the fastest velocity, which can transfer information, short pulses of light are ideal probe signals for fast-moving satellites. Since the satellites move on ellipses with the geocenter at one of the two focal points, sampling their orbits by laser ranging allows the center of mass of the Earth to be established with subcentimeter accuracy. So the geocenter conveniently became the origin of the terrestrial reference frame. The exact knowledge of the satellite orbit at the time of observation of the satellite as the geodetic marker allowed the exact positioning of the observing station on the ground.

Satellite laser ranging was not the only technique that was developed at the dawn of the space age. Another very important technology is Very Long Baseline Interferometry (VLBI). It was born when scientists started to realize that the Universe was also full of microwave radiation. There was not only the light of the stars, but also the radiation of supermassive black holes at the perimeter of the observable Universe. Interferometry of electromagnetic waves is a very powerful measurement technique. It reveals minute changes in the phase between two periodic signals with high resolution. The shorter the wavelength, the higher the measurement resolution. Interferometric measurements of two orthogonal arms of a Michelson type of interferometer provided measurements of gravitational waves for the first time in 2015, something that Albert Einstein thought would never be possible. To understand the involved sensor resolution, the change in distance compares to adding the diameter of a hydrogen atom to the mean Earth–Moon distance. By monitoring the same radio source in the sky and correlating the recordings, scientists can utilize microwaves with a wavelength of a few centimeters over distances of several thousand kilometers to determine accurate baselines between observing radio telescopes, now with uncertainties of around 1 cm over global distances. The suitability of this method was first demonstrated in April 1967 for the Dominion Radio Astrophysical Observatory and the Algonquin Radio Observatory, 3074 km apart. Some years later and with more telescopes involved, this technique provided the quantitative evidence for the crustal plate motion postulated by Alfred Wegner nearly 50 years before.

While SLR and VLBI provide several important points of reference on different locations and different continental plates, the Global Navigation Satellite Systems (GNSS) provide the necessary densification and the navigation functionality our modern society depends on. SLR and VLBI have the instrumental complexity placed on the ground. Therefore there are only a small number of stations. For the GNSS the complexity is in the space segment. GNSS satellites are no longer simple mathematical point masses, making up a geometric reference. They contain transmitters, highly accurate clocks and process codes. The detectors at the ground segment are fairly simple in comparison. That makes them cheap, and the concept inherently allows to operate many million of such receivers at the same time. When collocated with VLBI and SLR, the systematic errors of all of these techniques become accessible and can be reduced. The local survey between the reference points of the different systems provides a tie among the techniques and to the ground. This is an important step in the process of forming the terrestrial reference frame.

This book takes a technical approach and is intended for the reader with an interest in space geodesy, who wishes to understand the underlying technical concepts. It expects some basic technical interest and understanding, but certainly no specialized background. We, the authors, invite you to a technical tour around the toolbox of space geodesy. We are aware of the necessary fine balance between detail and generalization. For readers with deeper insight and special interest, we offer a lot of specific details, augmented by a list of references for further reading.

For the student of space geodesy with no particular physical background, we hope that we managed to provide sufficient background in an understandable way to obtain access to the basic technical ideas behind these important measurement concepts. We start with a general introduction to the common practice of physical experiments. The established techniques in space geodesy utilize time of flight measurements, therefore chapter 2 introduces the various clock concepts and the why, when and how they are used. Then we turn to the various optical ranging applications in the chapters 3–5, including optical time transfer and transponder applications. Optical interferometry is a powerful measurement technique, which can be use for inertial rotation sensing. Exploiting the Sagnac effect, based on optical traveling wave resonators, introduces a novel approach to accurate and highly resolved Earth rotation sensing. We put this into perspective with the small measurement quantities, that space geodesy explores. The reader has the opportunity to experience the incredible resolution that optical interferometry provides.

Space geodesy is complex, and while the underlying technical concepts are fairly simple, the extremely high resolution and stability that the network of stations has to provide is a great challenge. There are many very subtle geophysical processes involved that have to be untangled before a consistent result emerges. The book comes in two parts. Volume 1 prepares the basics and concentrates on the optical technologies and is written by Ulrich Schreiber. Volume 2 deals with the microwave technologies and explains GNSS and VLBI, before it looks at the propagation of microwaves in the Earth atmosphere. This part is written by Urs Hugentobler. Taken together, these books cover a lot of ground. None of the individual technologies covered in this book can provide the complete parameter set for the reference frames of space geodesy alone. The combination of all of the techniques is essential. We hope that our chosen approach helps to provide access to the complex machinery behind the scenes. We believe, that a solid technical understanding is a prerequisite for space geodesy.

Munich, 15 July 2025: Ulrich Schreiber, Urs Hugentobler

Author biography

Karl Ulrich Schreiber

Ulrich Schreiber studied physics at the 'III. Physikalisches Institut' of the Georg August Universität in Göttingen and graduated in 1984 on dielectric relaxation spectroscopy. He obtained his doctoral degree on ultrasonic absorption spectroscopy in 1988. Since 1988 he has worked for the Research Unit Satellite Geodesy at the Technical University of Munich in the field of space geodesy. He designed and managed the Wettzell Laser Ranging System on the Geodetic Observatory Wettzell. In 1999 completed his habilitation on Sagnac spectroscopy by the design of a large monolithic ring laser structure for applications in space geodesy. Another field of interest is accurate optical time transfer.

Ulrich Schreiber
Technical University of Munich
Research Unit Satellite Geodesy
Arcisstr. 21
80333 München
Germany
email: ulrich.schreiber@tum.de

Abbreviations

λ	Wavelength (m)
c	Velocity of light: 299 792 458 m s^{-1}
f	Frequency in units of Hz
G	Gravitational constant: $6.674\ 08 \times 10^{-11}$ N m^2 kg^{-2}
h	Planck constant: $h = 6.626\ 070\ 15 \times 10^{-34}$ J Hz^{-1}
K	Kelvin
k_B	Boltzmann constant: $k = 1.380\ 648\ 8(13) \times 10^{-23}$ J K^{-1}
μK	Micro-Kelvin
ps	Picosecond
ns	Nanosecond
μs	Microsecond
fs	Femtosecond
μas	Micro arc second
pm	Picometer
ppm	Part per million
ppb	Part per billion
rf	Radio frequency
ACES	Atomic Clock Ensemble in Space
alt-az	Altitude-azimuth
A/D	Analog-to-digital converter
AOM	Acustooptic modulator
APD	Avalanche photo diode
BC	Before Christ
BiPM	Bureau International des Poids et Mesures
CCR	Cube corner reflector
CCW	Counter-clockwise
CW	Clockwise
DORIS	Doppler Orbitography and Ranging Integrated by Satellite
EAL	Echelle Atomic Libre
ELT	European Laser Timing
ET	Event timer
ENSO	El Nino Southern Oscillation
GGAO	Goddard Geophysical and Astronomical Observatory
GNSS	Global Navigation Satellite Systems
GOPX	Galileo Optical Experiment
GPS	Global Positioning System
HeNe	Helium Neon gas laser
HEO	High Earth Orbiter
ICRF	International Celestial Reference Frame
IERS	International Earth Rotation and Reference System Service
IGS	International GNSS Service
IMU	Inertial measurement unit
ISS	International Space Station
ITRF	International Terrestrial Reference Frame
JD	Julian date
LEO	Low Earth Orbiter
LOLA	Lunar Orbiter Laser Altimeter

LRO	Lunar Reconnaissance Orbiter
mJD	Modified Julian date
MGS	Mars Global Surveyor
MOLA	Mars Orbiter Laser Altimeter
MOT	Magnetooptic trap
MWL	Microwave Link
NMI	National Metrology Institution
LLR	Lunar Laser Ranging
LoD	Length of day
PG	Pulse generator
PPS	Pulse per second time markers
PTB	Physikalisch Technische Bundesanstalt
Quasar	Quasi stellar object
SI units	International System of Units
SLR	Satellite Laser Ranging
SNR	Signal-to-noise ratio
SOS-W	Satellite Observing System Wettzell
T2L2	Time Transfer by Laser Link
TAI	International atomic time
TEM_{00}	Lowest-order laser mode with Gaussion profile
TWTT	Two-way time transfer
TWOTT	Two-way optical time transfer
UHV	Ultra-high vacuum
USO	Ultra stable oscillator
UT	Universal time
UTC	Universal time, coordinated
VLBI	Very Long Baseline Interferometry
WLRS	Wettzell Laser Ranging System
WTCO	Wettzell multitechnique station fiducial

IOP Publishing

The Measurement Techniques of Space Geodesy

Optical technologies

Karl Ulrich Schreiber

Chapter 1

Introduction

The main measurement techniques of space geodesy are realized by *Satellite Laser Ranging* (SLR), *Very Long Baseline Interferometry* (VLBI), *Global Navigation Satellite Systems* (GNSS) and *Doppler Orbitography and Radiopositioning Integrated by Satellite* (DORIS). These are augmented by local surveys and auxiliary measurements, like atmospheric pressure, temperature, humidity and inertial rotation sensing. The observed objects, which are satellites, quasars, stars and the inertial space around us, are communicating with our points of observation (the observatories) via electromagnetic signals. This happens in a variety of ways. We may distinguish between a unilateral one-way signal exchange and a bidirectional two-way measurement process. Examples of the former are VLBI, GNSS and the methods of astrometry. For the latter we can name SLR, Lunar Laser Ranging (LLR), asynchronous transponder operations, as well as delay-compensated time and frequency transmissions, either over an optical fiber or through free space. Rotational motions have a particular interest for us. Since we are living on a rotating planet and the Universe around us appears to be either at rest or in a steady translational motion (which we would be unable to work out), we need to keep track of this motion. Why? From the point of view of our measurement techniques of space geodesy, our main interest, a reference frame fixed to the Earth, is obtained from measurement targets that are not rotating along with the Earth. In order to disentangle the motion of the ground station from the motion of the targets, we need to apply corrections for the rate of rotation of the Earth, which is not constant. Furthermore, we need to know the instantaneous orientation of the rotational axis of the Earth. The interferometric observation of very remote celestial objects reveals these small deviations from a constant rate of rotation. Not only that, the rotational motion of the Earth is also visible in the physical properties of optical laser cavities, just as the effect of gravity is having an influence on the tick rate of a clock. In this book we will walk you through the elementary aspects of the measurement techniques of space geodesy. You will find out that there are only a few very basic concepts involved and they persistently keep coming back in different shapes.

doi:10.1088/978-0-7503-2032-0ch1

The observation techniques in space geodesy are characterized by extremely high requirements for resolution and stability. At the same time, they have to cover a broad range of frequencies. Some of the processes observed change very quickly–within seconds. Others take years to centuries or even longer to materialize. The Earth's rotation is one example of a fast changing process. Crustal motion on the contrary is an example that acts over a long time. To become observable we must establish a communication between the object of interest and some reference against which we compare. The measurement techniques, listed in the beginning, have one thing in common. The objects we observe are all located in space, while our observation instruments are located on the surface of the Earth. Object and instrument are communicating by electromagnetic radiation, either in the microwave domain or in the optical domain. All the techniques employ one property without saying so. The signals used for the interaction are traveling at the speed of light c, which is the highest possible velocity at which information can travel. The most important feature, however, is that the velocity of light is constant. Again, this is not stated anywhere, but it is the most important feature of all and this is what all the techniques have in common. Apart from that, the techniques are very diverse. They use ultrashort pulses or codes. They measure time of flight or generate interferograms. Some use satellites as targets, others use deep space targets, namely quasi stellar objects, known as quasars. Last but not least, some of the applied techniques are probing the local inertial space.

The periodic sinusoidal variation over time is an important feature of the measurement signals, which are employed here. This is actually no surprise. We quantify that by assigning a frequency value to it. The unit Hertz (Hz) is defined as a unit for the number of oscillations per second, 1 Hz = 1/s. All of the space geodetic techniques essentially perform a measurement of time in the end. The notion of time in the sense of physics is an ordering parameter, which separates events into a 'before' and 'after'. Time has a very important property. It is the SI unit, which can be established with the highest resolution of all. In this book we attempt to provide an overview over the toolbox of space geodesy from an experimental point of view. We wish to provide some insight into features and technical details of all these techniques. At the same time, we do not expect that the reader is deeply acquainted with physics or electrical techniques. We wish to introduce the concepts of the measurement techniques of space geodesy in a way that also the non-technical reader gets a solid idea of what is behind these concepts, their strengths and their limitations. We wish to show why there is no single perfect tool for this purpose. Naturally we will be moving along a fine line between detail and simplification and we hope that we have managed that well. Where helpful, we are pointing to further reading for the expert.

1.1 The basic concept of performing measurements

Over the centuries the natural sciences have developed a willful way to progress knowledge and to maintain consistency at the same time. Since it is impossible to prove something definitively right, they follow the opposite way. A single contradiction is enough to proof something wrong or at least to point to a limited applicability. Therefore the system works in a reversed way. When a new model is developed, based on earlier results and compatible with the current understanding, it immediately becomes subject to intense testing, even by the authors. In other words, there is a lot of effort going into shooting the new hypothesis down. The longer it survives, the better it is. In this process of testing new ideas for lasting validity, we must never forget that we do not know the true reality of our physical world. We only have a model of it available to us. The theory of relativity, for example, has substituted Newton's notion of absolute time and space. This does not make Newton's mechanics wrong. It only limits the applicability of this model to situations at low velocities that are compatible with our daily experience. We also do not expect that the general relativity model will be the final one. While it is just a model, it is a model with exceptional quality that has stood the test of time for more than 100 years. Nevertheless, there is no doubt that it will be substituted one day, however long this may take. At this point it is interesting to note that the measurement techniques in general are still a fair way behind the theory. They are only slowly catching up. Today we are still facing a situation, where an experimental test of general relativity is a major achievement and we do not even see a measurement sensitivity, that would allow us to prefer one alternative model over its competitors. Space geodesy is one of the disciplines, which is engaged in tests of general relativity. LLR for example is testing the Equivalence principle [3] and has accrued a time series of observations, spanning more than 50 years. As the time series is growing, the results are getting better. At this point in time, there is still no deviation from the predictions. The formal error in the calculations is still too large to draw a definitive conclusion.

The quality of a model in general can be rated by its ability to make predictions. The better these predictions are, the better the model is. If there are two or more competing models for a particular scientific question, the one that requires the fewest assumptions is usually considered to be the better one. This is the reason why scientists request *'beauty'*, *'simplicity'* and an *'aesthetic appeal'* from models. Since we cannot easily tell whether it is right or wrong in the absence of rigorous experimental tests, we only can look for some qualifying indicators for the case of competing ideas. Within a model, we can use already proven methods in order to arrive at further consistent predictions. When the ends do not meet eventually, then we have to rethink the entire process. It looks a bit like trial and error and to some extent that's what it is. In the end, it is the consistency that matters the most. Scientific progress usually follows a four-step approach:

1. Formulation of a hypothesis.
2. Experiment or calculations in accordance with approved practices including reasoning in consistent lines of causality.
3. Determination of an error budget including a critical review of the applied methods.
4. At this point, the hypothesis either survives or is shot down.

Let us look at an example to make this practice clearer. For illustration, we wind our calendar back by about 2000 years to the times of Eratosthenes of Cyrene. Let us think about the question, whether the Earth is flat or a sphere, within the technical possibilities of the time. Egypt of the time was very much concentrated along the river Nile, which happens to flow from south to north in a fairly straight way. Traveling from Alexandria to Aswan along the river covers a distance of more than 800 km. Looking at the stars, while traveling along that route would reveal that some stars in the north would disappear, while new ones would appear in the south. So this observation would contradict a flat earth, but is it a sphere? As we have seen from these first observations, we have a good reason to believe that we are living on a spherical earth. Now let us set up our hypothesis:

The Earth is spherical and we wish to establish the radius of this sphere.

This brings us to the experiment. How could we explore this question utilizing the available means in Egypt of the time? From today's perspective, the task of determining the radius of the Earth is straightforward and simple, but in the Egypt of 2000 years ago, it definitively required a genius to think this through. Figure 1.1 shows a sketch of the experimental layout. From the perception of ancient Egypt, there are two important prerequisites required. First of all, we need the insight, that the Sun is far enough away, that the wavefront of the solar radiation at the Earth can be considered as parallel. This is indicated by the arrows in figure 1.1. Furthermore, we need a good measurement of ΔL, which is a distance larger than 800 km. The experiment itself was simple. We set a pole upright with respect to local g on the ground at Alexandria and another similar one in Aswan. Some sources report of a vertical shaft instead of the pole, but this does not change the essence of the experiment. Around mid-summer, the Sun is at the zenith in Aswan and the shadow of either the pole or in the bottom of the shaft disappears at noon. In Alexandria one then measures a value of the shadow angle of $\gamma = 7.\ 4^{\circ}$ at the same day. With γ and ΔL measured, the radius (R) of the Earth can be calculated as

$$R = \frac{\Delta L}{\gamma}. \tag{1.1}$$

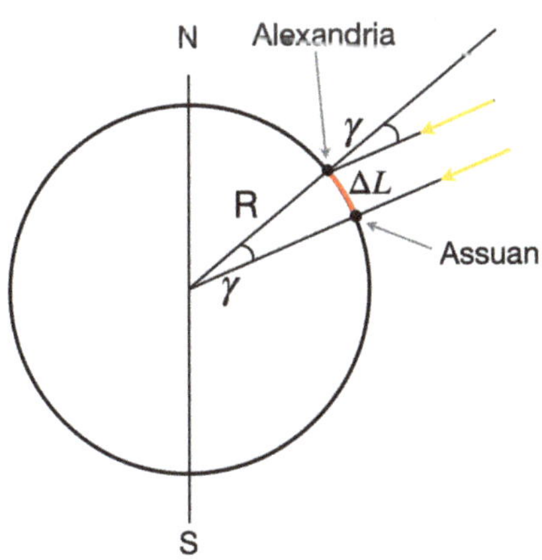

Figure 1.1. The experimental layout of the proof of concept (adapted from [8]).

It is remarkable that the value for the radius of the Earth, determined by Eratosthenes turned out to be correct within 2% of the mean radius and it is very likely, that this is not so by accident. Approximately 100 years later, another attempt was carried out by Posidonius. He wanted to reduce potential errors by making ΔL a straight line. So he chose the distance between Alexandria and the Isle of Rhodos. His results were out by 11% and this brings us to item #3 in our list. Planning an experiment and performing it are essential, but this must necessarily be accompanied by a critical review of all the involved aspects.

Let us return to Eratosthenes. In his day, Egypt already had a cadaster and distances between cities and important locations were routinely reestablished by step-counting people [8]. This happened not just once, it happened repeatedly. So we can assume that the determination of ΔL was surely not entirely correct, but very reasonably consistent. If we assign the entire error on the distance, it would come out as an uncertainty of ±16 km. The cities of Aswan and Alexandria are not exactly on the same meridian. Does that matter for the measurements and in the error budget? I think not. The line between the two cities is only 3° out of a straight north–south line and this error enters as a cosine, which remains close to 1. This error would have an effect of 2 km on ΔL, which in the full context of this experiment is too small to matter. For the purpose of this measurement it is not necessary to obtain the angle from the shadow pointers at the same time. In fact, once you know at which days in the year the Sun is in the zenith in Aswan, you do not even need to do that measurement there. It would be enough to only measure in Alexandria. The valid angle for the purpose is when the length of the shadow is the shortest. That means, one does not even require a clock for that matter. Our experiment requires that the shadow pointer is vertical in order to provide the correct angle. This condition is easily met with a plumb bob. An error of 2% entirely caused by the angle reading, would have caused a deviation from the real value by ±0.15°. We see, the angle needs a careful measurement and we can assume that it certainly has been measured more than once. So why did the measurements from Posidonius turned out a larger error? The distances and the offset angle in both measurement campaigns were fairly similar. The inherent advantage of using a straight line across the Mediterranean was compromised by the problem of measuring distances on the sea correctly. While it is in principle possible to measure the velocity of the boat relative to the water, ocean currents or wind drifts could not be measured, which increased the measurement error considerably. Posidonius also did not have the luxury that the Sun was in the zenith at any time of the year.

Based on a clever choice of the experiment and a good understanding of the boundary conditions made it possible to determine the radius of the Earth with impressive accuracy already more than 2000 years ago. From this episode we take the message that the necessity to constantly reflect over the experimental components is not an exception; it is our standard experimental requirement. We are always in the same situation. There is always the challenge to identify critical parameters and often it will result in the need to rearrange our experiment. Some aspects that have not shown up in the beginning, may turn out to be critical showstoppers. New inventions may change a technical concept quickly and dramatically. Here is an example from recent times. When satellites became available from 1957 onward and the concept of a laser appeared in 1960, the two technical components very quickly morphed into a new measurement technique, namely SLR. Together with VLBI it completely changed the way of taking global geodetic measurements. When Alfred Wegener formulated his famous plate tectonic theory in 1925, nobody knew how fast these plates were moving. Were they drifting apart at a rate of mm per year or several meters and to some was it existing at all? Experimental proof was provided by space geodesy sometime around the year 1976, derived from SLR and VLBI measurements. It turned out to be on the scale of several cm per year and the continental plates did not move everywhere at the same rate. Alfred Wegener had set up his hypothesis based on fossils found in Africa and South America and the coastline contours of the continents.

Now let us construct another example, even if it is a bit exotic and far-fetched. The hypothesis is a proposal for measuring plate tectonics without SLR or VLBI. We assume that we have two meridian observation stations, one on each plate. After looking at the coastline contours, we take it as an indication, that our experiment is testing an east–west motion. We measure the transit of the Sun on each observatory on a daily basis over a long time and we time tag these transits against the UTC time scale. Figure 1.2 exhibits a simple sketch of our thought experiment. Since the Earth is rotating uniformly, one would expect that the time between the meridian transits in both places will slowly increase. This should eventually give qualitatively a diagram like the one shown in figure 1.3. From what we know about plate tectonics this would be a very plausible experiment. Is it really? I think we need to take a closer look at the abundant pitfalls in experimental physics. In situations like this, it is important and good practice to set the envisaged result and the available resolution of the employed measurement techniques into proportion. It is also important to check if all the assumptions made in the design of the experiment are valid. Conceptually, we measure time intervals in this thought experiment. So let us do some numbers. Between the two stations we have, say, 8000 km. From one transit to

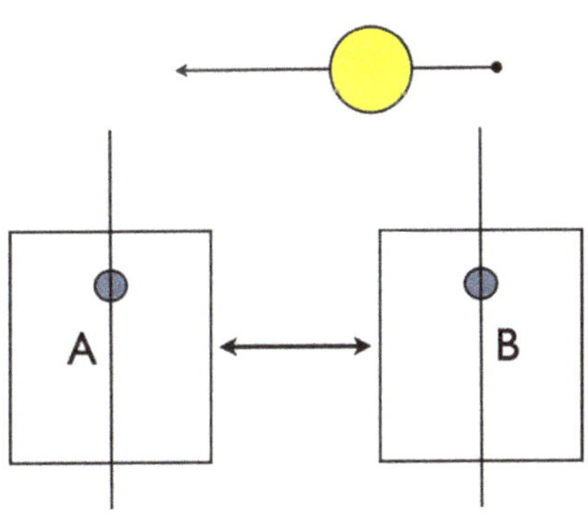

Figure 1.2. Graphical illustration of our thought experiment. The areas A and B represent the continents, which are expected to move apart. The uniform Earth rotation lets the Sun pass with a constant rate between the two meridian transit stations.

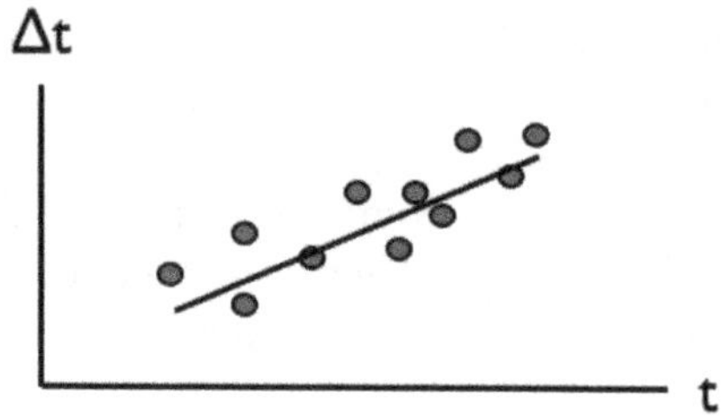

Figure 1.3. Here is qualitatively what we would expect from the experiment. The linear trend would represent the desired rate of continental drift.

the next we have 86 400 seconds, which is the length of the solar day. We would be lucky to determine the moment when the Sun touches the meridian line to 1 second of accuracy, since the boundary of the Sun is not so sharply defined. The length of the equator is 40 000 km. This divided by 86 400 gives 463 m s^{-1}. At mid-latitude of say 41° this would provide a measurement resolution of about 300 m. When we compare this to the actual rate of the drift rate of the plate tectonics of 1.8 cm year^{-1}, this relationship is sobering. But this is not all. For our experiment we have assumed that the Sun is moving at a constant rate. However, the Earth is not in an ideal circular orbit. So here we hit another snag. Furthermore, we also assumed a constant rotation rate of the Earth, which also is not a given. The relative distance d between both stations would change by

$$\frac{\delta d}{d} = \frac{0.018\ \text{m}}{8\,000\,000\ \text{m}} \approx 2 \times 10^{-9} \tag{1.2}$$

over one full year! What a tall demand. Values like the one obtained here in equation (1.2) are fairly typical for the measurement techniques in space geodesy. This enormous discrepancy between the design of our thought experiment and reality is the final nail to the coffin of our hypothesis and this is not unusual. There are many more ideas, that have not survived than those ideas that prevail.

1.2 Error sources in performing measurements

The measurement of a physical quantity X always comes with an error. First, we have the problem that the measurement effort perturbs the experiment. Then we find that the measurement itself is not perfect either. We distinguish between:

$$\text{absolute errors} \qquad \delta x = X_{\text{mess}} - X \tag{1.3}$$

$$\text{relative errors} \qquad x_{\text{rel}} = \frac{\Delta x}{X} \tag{1.4}$$

We distinguish between several categories of errors. **Objective errors** point to an imperfect experimental setting or environmental problems during the experiment. A change in ambient temperature may compromise the function of the measurement device. Then there are **subjective errors**. These usually occur when a measurement device is set to the wrong range or an inappropriate function. A **constant error** causes an offset to the measurement. An error in reading the time is a classical example. On a more subtle basis are reading errors when the needle of a scale is read under an angle (parallax). This falls under the category **systematic error**. A particularly nasty group of errors are the **random errors**. We encounter them fairly often and they are particularly difficult, because they are often not repeatable. For example, electrical potential fluctuations when voltages are taken often are contaminated by this error. This kind of error usually is an indication for an inadequately designed experimental setup. Usually it takes a very long time to find this out, and a tedious effort is often required to get rid of them.

If the experiment is composed of several parameters, such errors may accumulate. In such cases we have to apply the **law of error propagation**

$$\Delta x = \sum_{i=1}^{k} \left| \frac{\partial f}{\partial x_i} \right| \Delta x_i. \tag{1.5}$$

In the strict form, as presented in equation (1.5), all error contributions are summed up with the result that the estimated error takes the maximum value. In the sole presence of systematic errors, this law is usually applicable, provided these errors are small with respect to the measurement quantity. By repeating a measurement with random errors often, we can reduce the effect of the errors. We establish the **mean value** of a measurement series as

$$\bar{x} = \frac{1}{n} \sum_{i=1}^{n} x_i. \tag{1.6}$$

The measurement series {1, 2, 4, 5} and {2.7, 3.0, 3.1, 3.2} provide the same mean value of equation (1.6), namely $\bar{x} = 3$, although they are significantly different. The first series is less sharply defined than the latter series. Therefore, we need an additional quantity to establish the spread about the mean value. This quantity is the **variance** and for more than one measurement value it is defined as

$$\sigma^2 = \frac{1}{n-1} \sum_{i=1}^{n} (x_i - \bar{x})^2. \tag{1.7}$$

The positive square root of the variance defines the **standard deviation** and has the great advantage that it has the same unit as the measurement values. Provided a set of measurements is only contaminated by random errors, we will find that they exhibit a **normal distribution**. Another name for that is **Gaussian distribution** and it is represented by

$$f(x) = \frac{1}{\sigma\sqrt{2\pi}} e^{-\frac{(x-\mu)^2}{2\sigma^2}}. \tag{1.8}$$

Figure 1.4 represents the normal distribution from equation (1.8) for the values σ=0.25, σ=0.5 and σ=1. We note that 68% of all measurement values fall into the interval $\pm\sigma$ and 95.45% of the measurements are found between the values $\pm 2\sigma$. Because of its importance we note this again. The mean value and the standard

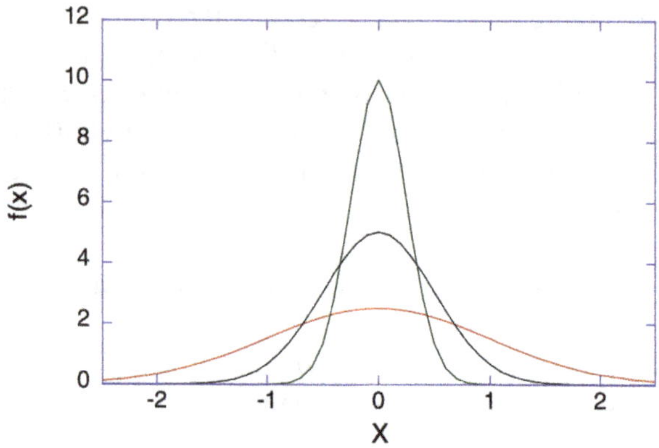

Figure 1.4. The normal distribution for three different values of σ, namely 0.25, 0.5 and 1. The smaller the σ-value, the sharper is the distribution.

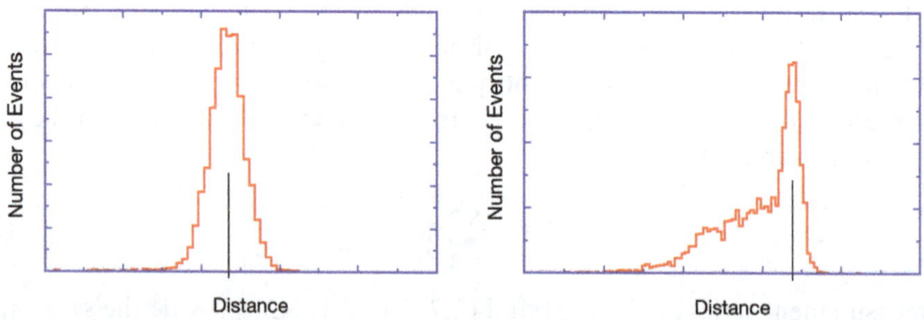

Figure 1.5. Example of a repeated laser ranging measurement, where the echoes are normal distributed (left). The true distance is indicated with the vertical line and the computed mean value and the standard deviation will be correct. The other relative extreme example (right) shows measurements experiencing a systematic error. The true distance is not available from a simple statistical analysis.

deviation **only** applies to measurements that exhibit a normal distribution. As soon as systematic errors are part of the measurements, we find that the distribution is skewed and the estimated values for $\bar{x}$ and σ are not correct. We illustrate this with an example. In figure 1.5 there are two histograms from a laser ranging application shown. The diagram on the left side, shows a flawless observed laser ranging measurement. The echoes are normal distributed and the mean value and standard deviation come out at the true values of the experiment. The diagram on the right side shows a compromised measurement. Due to some light leakage, there is a systematic error to one side in the histogram. While the true range is indicated by the vertical lines in both charts, the calculation of the standard deviation and the mean value would not reflect a reliable result. In fact, both of these quantities would turn out as meaningless. While this example here is rather extreme, we will deal with significantly less obvious cases most of the time. So we need to be aware of this issue. Applied to the technique of laser ranging we can summarize this paragraph in the following way. For an often repeated calibration measurement, each single data point deviates on average as $\Delta\bar{x} = \sigma/\sqrt{n}$ from the mean value. Random errors, like the influence from the pulse width of the laser, the bandwidth of the detector or the granularity of the timer increase the scatter of the measurement, but still produce the correct mean value. Systematic errors, however, like an intensity-dependent detector delay, are causing a bias and this leads to incorrect ranges. While this is here formulated for the laser ranging application, it is still a universal property, which only differs in the describing terminology.

We have presented the strict error propagation law above. A somewhat softer formulation is usually more applicable for all practical purposes. The **Gaussian error propagation law** makes use of the fact that some of the errors may enter with a different sign, so that they neutralize each other in parts. It is:

$$\Delta x = \sqrt{\sum_{i=1}^{k} \left(\frac{\partial f}{\partial x_i}\right)^2 \sigma_i^2} \tag{1.9}$$

A simple example illustrates equation (1.9). The value of two resistors has been measured repeatedly and the values $R_1 = 100 \pm 0.8\ \Omega$ and $R_2 = 200 \pm 1.0\ \Omega$ were obtained. What is the total resistance when we put them in series and how large is the error ($R = R_1 + R_2 = 300\ \Omega$)? For the Gaussian error propagation law we need the partial derivatives, namely $\partial R/\partial R_1 = 1$ and $\partial R/\partial R_2 = 1$. With $\sigma_1 = 0.8$ and $\sigma_2 = 1.0$ we obtain the mean error as

$$\Delta R = \sqrt{\left(\frac{\partial R}{\partial R_1}\right)^2 \sigma_1^2 + \left(\frac{\partial R}{\partial R_2}\right)^2 \sigma_2^2} = \sqrt{0.8^2 + 1.0^2} = 1.3 \tag{1.10}$$

The combined resistance of the two resistors in series is $R = 300 \pm 1.3\ \Omega$ instead of $R = 300 \pm 1.8\ \Omega$ as the strict error propagation law would provide. A wider treatment of measurement errors are found e.g. in [7].

1.3 Time series analysis

The estimation of mean values and the standard deviation are suitable tools for the evaluation of static processes. Space geodesy and metrology in general feature situations, where things are changing dynamically. A better way to phrase this is that the objects under test are drifting. Take for example an optical resonator and watch the output frequency over some time. It will certainly drift off in some way, responding to changes in ambient temperature. Figure 1.6 represents the observation of a Sagnac interferometer. It should produce a beat note, which only depends on the experienced rotation rate of the Earth (refer to chapter 6 for details), which at this level of resolution can be considered constant. However, this is not what we observe. How do we deal with such an effect and get some extra information out of the measurement at the same time? A simple statistical analysis is not applicable, since we have some systematic effect on the data. Looking at the standard deviation would also not reveal anything about the underlying noise processes. However, this is what we are interested in and a proper characterization of oscillators, clocks or time scales is a very common problem. The actual frequency is not of concern in this

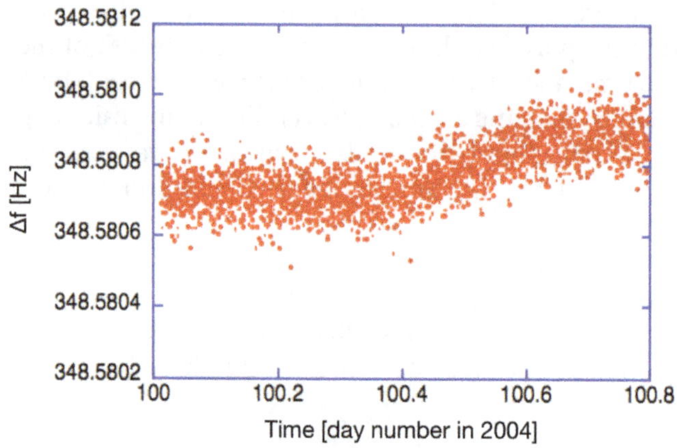

Figure 1.6. Example of the output frequency of an optical Sagnac interferometer. The dataset shows evidence of a frequency drift.

context, but the relative instability is a key figure. Therefore, we normalize the frequency fluctuations $y(t)$ to their nominal value ν_0

$$y(t) = \frac{(\nu(t) - \nu_0)}{\nu_0} \tag{1.11}$$

The normalized frequency can be measured directly with a time interval counter, for example, in a clock comparison. It makes up the rate of change between a device under test and a reference source within the chosen length of the measurement interval. (Example: The difference of 1 ns after 10 s of integration time corresponds to a deviation of 10^{-10}.) Since we use a time interval of the duration τ, we obtain

$$\bar{y}(t) = \frac{(x(t+\tau) - x(t))}{\tau} \quad \text{and} \quad x(t) = t_2 - t_1. \tag{1.12}$$

The quantity $y(t)$ takes the role of the instantaneous frequency, but strictly speaking, this is not the case, because it cannot be generally observed as a function of t. It rather represents the mean value over a fixed interval Δt, which, however, has the same length as the parameter τ in the Allan variance. (We say $y(t)$, but actually mean $\bar{y}(t)$.) In a similar way we can express the difference of the phase in a clock comparison.

$$x(t) = \frac{\varphi(t)}{2\pi\nu_0} \tag{1.13}$$

Since frequency is the time derivative of the phase, we obtain

$$y(t) = \frac{\mathrm{d}x(t)}{\mathrm{d}t} \equiv \dot{x}(t) \quad \text{and} \quad x(t) = \int_0^t y(t')\mathrm{d}t' \tag{1.14}$$

Figure 1.7 illustrates the relationship between $x(t)$ and $y(t)$ in the top part (a) and the forming of the differences in the bottom part (b). It is desirable to switch from the

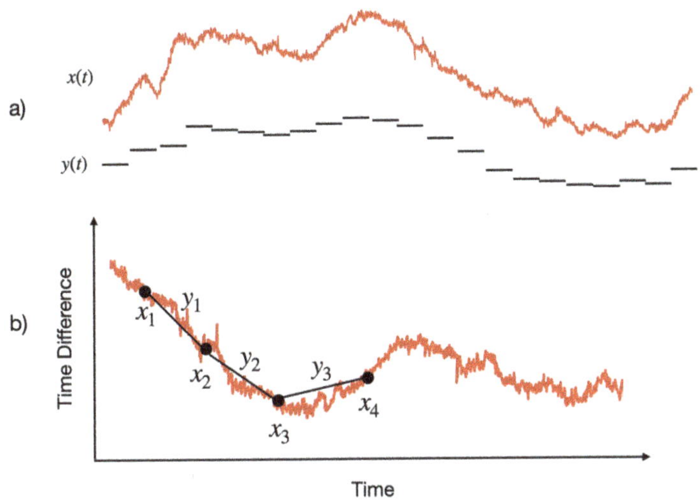

Figure 1.7. Graphical illustration of forming the time differences $x(t)$ and the frequency differences $y(t)$.

absolute units of the variance in equation (1.8) to relative units as a function of the averaging time τ. This is known as the **Allan variance** [1, 6].

$$\sigma_y^2(\tau) = \frac{1}{2(n-1)} \sum_{i=1}^{n-1} \left(\frac{x_{i+1} - x_i}{\tau} \right)^2 \tag{1.15}$$

In the absence of systematic errors we have a very typical situation for precision measurement systems. The short-term behavior is characterized by random errors, while in the long-term they develop a drift. Usually we control the short-term by system calibrations. However, for the long-term we need to explore what is causing the observed trend. Equation (1.15) is very helpful in this respect, because different drift causing processes develop distinctly different slopes, when the Allan variance is represented in a double logarithmic diagram as a function of the averaging parameter τ. The square root of the Allan variance is the **Allan deviation,** also frequently abbreviated as ADEV. The Allan deviation, similar to the standard deviation exhibits the same units as the measurement. From an ADEV plot we can identify different noise processes by their characteristic slope of the general form $\sigma_A = \tau^x$. A few of these processes are listed in table 1.1. The cause of the unfavorable oscillator behavior can be directly read from the diagram when the ADEV is displayed [5]. A ring laser gyroscope can be considered as an optical oscillator in this respect. Figure 1.8 illustrates a practical application of the ADEV. For integration times between 60 s and 1 h we find a slope for σ_A of -0.5, indicating white frequency noise. Between 1 and 3 h we obtain the best sensor sensitivity and it is disputable to state that we have reached the $1/f$ flicker floor here. In this diagram it rather looks like we see the transition from white frequency noise to a random walk process. The corresponding slope increases with σ_A of 0.5, before it turns round and continues to decrease again as white frequency noise. What was not known at the time of this

Table 1.1. Some typical drift causing process and their characteristic slope.

Behavior	Cause
$\sigma_A \propto \tau^{-1}$	**White phase noise** (Quantization noise, caused by the discrete nature of the sampling)
$\sigma_A \propto \tau^{-1}$	**White phase flicker noise** (Changing properties in the electronics of the apparatus)
$\sigma_A \propto \tau^{-0.5}$	**White frequency noise** (Noise from quantum processes like spontaneous emission)
$\sigma_A \propto \tau^{0}$	**Frequency flicker noise** This noise process shows 1/*f*-behavior
$\sigma_A \propto \tau^{0.5}$	**Frequency random walk** The system goes from one stable state into another stable state
$\sigma_A \propto \tau^{1}$	**Sensor Drift** This effect can be caused by ambient parameter changes

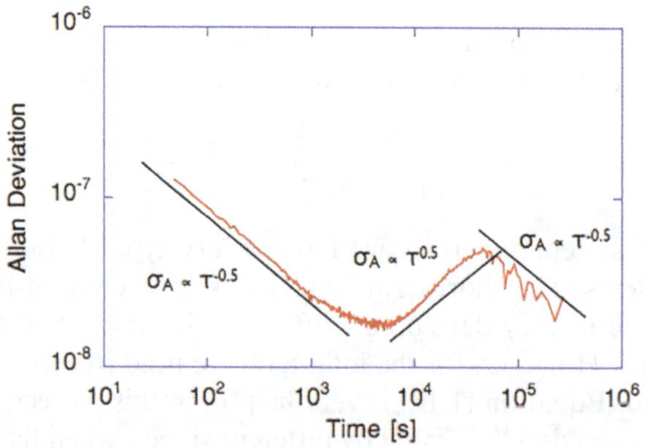

Figure 1.8. The calculated ADEV of the G ring laser gyro at the Geodetic Observatory Wettzell.

measurement, that there is a geophysical set of processes with a set of frequencies with periods around one day, causing the hump in our ADEV analysis. Right here, we use this example despite its limited validity to illustrate the importance of the Allan deviation analysis. The ADEV will be discussed in more detail in sections 2.1.1 and 6.4.2.

Repeated measurements do not only provide the measurement errors, they also reveal erroneous measurements. These are measurements that do not contain a valid result. This can point to a defect in the measurement system or parts of it. It can also mean that the conditions for a measurement are not right. Under normal conditions instrumental defects will cause suspicion very quickly, when the observations are inconsistent or pure nonsense. However, the effect can also be very subtle, in particular when there is a failure in some minor device out of a larger setup. A classical problem occurs when several instruments in one setup need to be synchronized to a frequency reference and it turns out later, that one of them was still running on its internal clock instead of the external reference. It is therefore absolutely mandatory to be critical about every result that is obtained. Are the measurements in the right regime, both in terms of spectral content and with respect to the obtained values (e.g., intensities). Measurements in space geodesy can have a very small signal-to-noise ratio (SNR). Take, for example, a daytime observation of a high orbiting satellite. Satellite echoes create a response on the detector, but background daylight will also do that. The former is what we want, the latter is much more frequent. How do we tell them apart? Detections caused by daylight are distributed randomly all through the detection gate (for details look at section 3.2.7). They will have no fixed relationship to the range prediction, hence they do not show any order. Valid range measurements, however, have a fixed relationship to the respective predicted distance. They will accumulate with a distinct pattern. In other cases one can exclude possible detections when the amplitude is larger or smaller than a given threshold.

In this section we have briefly touched on some of the tools and considerations that are essential in space geodesy as well as in other applications. We have realized that improved resolution and stability are essential requirements. The single most important contribution of space geodesy to science is a consistent highly resolved global reference frame with an incredible long-term stability [4]. A relative measurement resolution at the level of 10^{-9} is the goal. This corresponds to a range resolution of 1 cm over a distance of 10 000 km, or an angular resolution of less than 100 μas. This is a serious challenge and we need to take a clear distinction between **precision** and **accuracy**. The former of these two qualities expresses the amount at which our measurements scatter, while the latter is far more difficult to establish, because the respective quantities need to be without systematic errors, and this is much harder to achieve. An advanced treatment of mathematical methods in the physical science is found in [2].

References

[1] Allan D W 1966 Statistics of atomic frequency standards *Proc. IEEE* **54** 221

[2] Boas M L 2005 *Mathematical Methods in the Physical Sciences* (Wiley)

[3] Müller J, Murphy T, Schreiber K U, Shelus P J, Torre J M, Williams J G, Boggs D H, Bouquillon S, Bourgoin A and Hofmann F 2019 Lunar Laser Ranging: a tool for general relativity, lunar geophysics and Earth science *J. Geod.* **93** 2195–210

[4] Pearlman M and Plag H P 2009 *Global Geodetic Observing System: Meeting the Requirements of a Global Society on a Changing Planet in 2020* (Berlin: Springer)

[5] Riley W J 1995 A test suite for the calculation of time domain frequency stability *Proc. 1995 IEEE Frequency Control Symp.* pp 360–6

[6] Sullivan D B, Allan D W, Howe D A and Walls F L 2001 Characterization of clocks and oscillators *NIST Technical Note* **1337** 1–357

[7] Topping J 1972 *Errors of Observation and their Treatment* (Dordrecht: Springer)

[8] Torge W 1980 *Geodesy* (Berlin: de Gruyter)

Chapter 2

Time and frequency

All the established measurement techniques of space geodesy are based on the measurements of time or time intervals. Therefore, it is necessary to look at high precision clocks as the essential and very fundamental measurement tool. Although we are interested in the geometric relationship between the observed object and the observing instrument, it is the clocks that we are actually dealing with in several different ways. Clocks can be grouped in different categories, depending on the size, the weight, the robustness, the resolution, the accuracy and last but not least the stability. Satellites require different clocks than metrology institutions. They must be rather small, low in weight, robust and suitable for smooth operation in an adverse environment. Quite naturally such a clock would neither be accurate, nor would it be stable long term. On the contrary, the highly accurate and long-term stable clocks of a national metrology institution would not fit into a satellite, let alone be accurate under adverse operating conditions. In this chapter we look at time and frequency, two flavors of essentially the same thing, and the way they are utilized in space geodesy.

2.1 Historical background: a short history of time

Every one of us is used to clocks. We usually carry them around either on our wrist, around our neck or as an essential component of our mobile telephone. We are used to reading the time, but we do not know what time really is. There appears to be a fine line that separates the future from the past, two aspects of time that we can never reach. We are confined to live in the present, which is this apparently moving slider that converts a hazy future into a definitive past.

In physics time takes the role of an order parameter, which is useful to put physical events into sequence. A physical event can be the initial moment of a drop experiment, when a piece of mass starts to move or the final moment when the weight is hitting the bottom. Unlike the motion of an object in a three-dimensional space, which can move forward and backward, time is a strictly monotonous growing quantity. It helps us to distinguish a *before* from an *after* or a *simultaneous*. Physical equations that carry the

doi:10.1088/978-0-7503-2032-0ch2

parameter time are also meaningful, when time is entered as a negative value. However, the corresponding effects are never observed. While we can observe a glass object dropping from a shelf to the ground and breaking, there is no account of glass debris rising from the floor, reassembling the object as it moves up to finally sit on the shelf in one piece. The arrow of time is strictly pointing in one direction only. Nothing else has ever been observed.

For our purpose in space geodesy we take the arrow of time for granted and we also do not speculate about time as such. In fact, we basically utilize a property of clocks, namely that they are able to separate the observed apparent flow of time into intervals of equal length. The most general function of a modern clock is that it runs an internal oscillator, which ideally produces a sinusoidal output signal of a distinct single frequency. The clockwork then allows us to count these oscillations and thus generate a time scale, which is strictly only valid for this particular clock. But what are the units? Over many centuries a small set of units of time has emerged as being useful for our daily lives. We have defined the duration of one hour, one minute and finally one second and their respective conversion rate, according to the needs of a technically developing society. There are 60 s in 1 min and 60 min in 1 h. 24 h that make up the entire length of one day. All this is given by definition. There is nothing in nature that would provide us with a perfect and stable natural time scale. A clock that displays, hours, minutes and seconds then must have a conversion factor applied, which transforms the cycle count of the clock oscillator to seconds. Figure 2.1 illustrates this in a very basic block diagram.

It is also very typical and common that even two clocks of the same type, which are built as identical as possible, are not running at exactly the same rate. Furthermore, a single clock will also run at different rates, when the environmental parameters change. We may summarize the properties of clocks in a few statements:

- Each type of clock generates its own time scale.
- The time scales of different clocks of the same type are not necessarily identical.

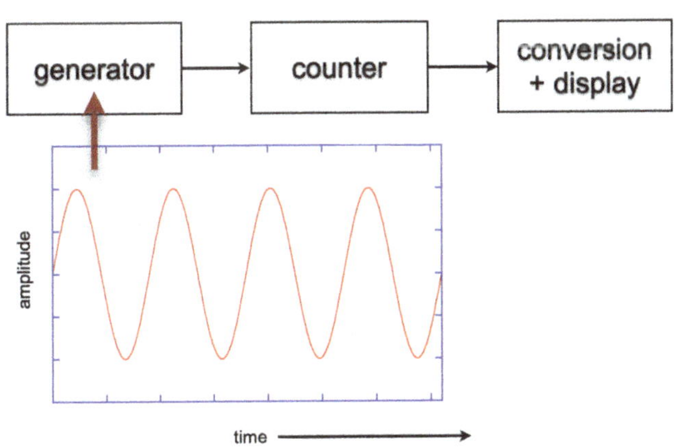

Figure 2.1. Block diagram of the basic clock functions.

- All type of clocks are based on the same laws of physics.
- The linearity of a time scale as well as the granularity (resolution) and the offsets relative to a defined origin are the important characteristics of a clock.

It is very obvious that the quality of a clock entirely depends on the generator in several ways. Higher frequencies of the generator give better resolution for the measurement of time as the clock accrues more cycles per unit time. The stability of the generator frequency ensures that the conversion factor remains constant and that means the clock does not drift. While historical clocks were only set up to count full cycles, modern electronic clocks can also evaluate the phase of the generator signal in order to resolve fractions of a cycle, which has the advantage of making the time scale realized by a particular clock more continuous. From these considerations we can define the actual resolution σ of a given clock by taking the uncertainty Δf of the generator frequency and dividing it by the operating frequency of the generator.

$$\sigma = \frac{\Delta f}{f} \tag{2.1}$$

Let us look at a few examples. Table 2.1 provides some examples of clock performances.

The table illustrates the advantage of going to progressively higher generator frequencies quite well. It also shows an impressive stability of the mechanical frequency defining system of the grandfather clock, despite the fact that the overall outcome of a loss of 12 min in one day is not all that great. The variability Δf of only 4.2 mHz on a purely mechanical system is indeed quite an achievement, but the slow oscillation frequency of the pendulum, on the other hand, degrades the performance significantly. Commercial clocks of this making were considered to be interpolators. They required a resynchronization once per day (typically at noon)[1]. A pendulum clock essentially represents a physical representation of a *harmonic oscillator* and it is said that Galileo Galilei (1564–1642) already observed that the period of oscillation

Table 2.1. Some examples to demonstrate the development of clock resolution

Clock type	f [Hz]	σ	Error in one day [s]
Grandfather clock	0.5	8.4×10^{-3}	726
Quartz	10^6	2×10^{-7}	0.02
Cesium	$\approx 9 \times 10^9$	2×10^{-13}	2×10^{-8}
Calcium	$\approx 4.5 \times 10^{14}$	6×10^{-18}	5×10^{-13}

[1] The record for a mechanical pendulum clock is the *Shortt* clock, which is nearly free of friction and runs in vacuum. These clocks only lost a few milliseconds per day.

(T) depends on the length of the rope (l), but not on the mass of the attached weight. With the gravitational acceleration g we obtain

$$T = 2\pi\sqrt{\frac{l}{g}}. \tag{2.2}$$

From a practical point of view, such an oscillator is not very suitable to be a clock generator. The period of swing is fairly long and friction quickly damps the oscillation. However, this example is very useful to study the harmonic oscillator. For small excursions of the weight at the bottom (in order to keep the system linear), we obtain the restoring force F as

$$F = -sx, \tag{2.3}$$

where x is the excursion and s the spring constant, which describes the stiffness of the system. The minus sign indicates that the force is a restoring force that counteracts the excursion. By setting $F = m\ddot{x}$ and performing some manipulations we obtain:

$$m\ddot{x} = -sx \tag{2.4}$$

$$m\ddot{x} + sx = 0 \tag{2.5}$$

$$\ddot{x} + \frac{s}{m}x = 0 \tag{2.6}$$

A comparison of the involved units shows $F = \frac{\text{kg m}}{s^2} \rightarrow \frac{\text{s}}{\text{m}} = 1/T^2 = \nu^2$. That means that the units of the coefficient in equation (2.6) correspond to a frequency. We will see shortly that we are dealing with a sinusoidal strictly periodic oscillation. Therefore, it is more appropriate to substitute the frequency with the angular frequency $\omega = 2\pi\nu$ and the expression for the oscillation period T becomes

$$T = \frac{1}{\omega} = \frac{1}{2\pi}\sqrt{\frac{m}{s}} \tag{2.7}$$

When we apply equation (2.7) to equation (2.6) we obtain

$$\ddot{x} + \omega^2 x = 0, \tag{2.8}$$

which is the differential equation for the **harmonic oscillator**. To solve this equation it is useful to make use of trigonometric functions.

$$x = A\cos(\omega t); \quad \dot{x} = -A\omega\sin(\omega t); \quad \ddot{x} = -A\omega^2\cos(\omega t) \tag{2.9}$$

Applied to equation (2.8) we obtain

$$\ddot{x} = -A\omega^2\cos(\omega t) = -\omega^2 x, \tag{2.10}$$

where A stands for the amplitude. The second solution is:

$$x = B\sin(\omega t); \quad \dot{x} = B\omega\cos(\omega t); \quad \ddot{x} = -B\omega^2\sin(\omega t) \tag{2.11}$$

Applied to equation (2.8) and setting the amplitude to B it gives

$$\ddot{x} = -B\omega^2 \sin(\omega t) = -\omega^2 x. \tag{2.12}$$

The superposition of both solutions

$$x = A\cos(\omega t) + B\sin(\omega t) \tag{2.13}$$

provides a general solution for the differential equation of the harmonic oscillator

$$\ddot{x} = -\omega^2(A\cos(\omega t) + B\sin(\omega t)) = -\omega^2 x. \tag{2.14}$$

The coefficients A and B correspond to values of x and $\dot{x}$ at a given time and for a fixed phase angle ϕ.

$$A = a\sin(\phi) \quad \text{and} \quad B = a\cos(\phi) \tag{2.15}$$

By taking the square of A and B we obtain a rather simple expression for the amplitude a, which is

$$A^2 + B^2 = a^2 \sin^2(\phi) + \cos^2(\phi) = a^2 \tag{2.16}$$

and

$$a = \sqrt{A^2 + B^2} \tag{2.17}$$

By the substitution of A and B into the general solution of the differential equation and the application of the theorem for the summation or subtraction of trigonometric functions ($\sin(\alpha \pm \beta) = \sin\alpha\cos\beta \pm \cos\alpha\sin\beta$), we obtain a simple representation for the state of motion of a harmonic oscillator.

$$x = a\sin(\phi)\cos(\omega t) + a\cos(\phi)\sin(\omega t) = a\sin(\omega t + \phi) \tag{2.18}$$

We obtain a sinusoidal pattern of motion over time. It can be viewed as the track, which is formed by a vector of length a (amplitude), rotating with constant angular velocity and progressing linearly along the time axis. The phase angle ϕ provides the initial start condition. Figure 2.2 illustrates the solution. There are a lot of harmonic oscillators available all around us. They can be found in mechanical systems, electronic circuitry, made from coils and capacitors, quartz crystals and also atomic

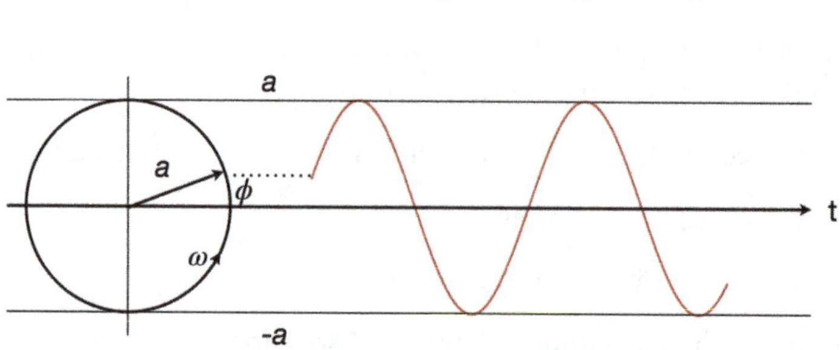

Figure 2.2. Graphical illustration of the track of motion of an harmonic oscillator. The vector with the amplitude value a rotates along the circle with the angular velocity ω, while the angle ϕ corresponds to the initial offset.

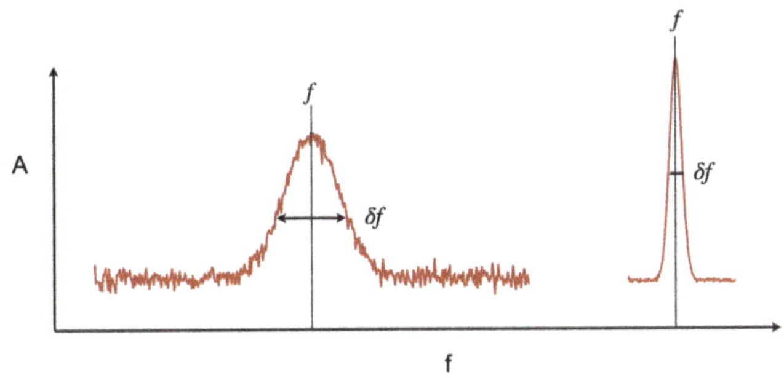

Figure 2.3. Illustration of the resonance of an oscillator of a low Q factor (left) and with high Q (right). A very sharp resonance means it has a high Q and only very few frequency components around the value of the peak value are contained in the output signal.

transitions. For the case of an electronic circuit the corresponding equation takes the form $L\ddot{q} + \frac{q}{C} = 0$ and $\omega = 1/\sqrt{LC}$. In this equation L indicates the inductance of a coil, C the capacitance and q the charge. The amplitude has a maximum at the resonance frequency. When the system is slightly detuned either above or below the resonance, the amplitude will drop. The further the system is detuned the lower the amplitude gets until nothing is left. In some cases this fall-off in frequency happens very rapidly and in other cases it takes much longer. It is therefore useful to define a quality factor Q, which characterizes a system. We define Q as

$$Q = \frac{f}{\delta f}, \tag{2.19}$$

where f is the oscillation frequency and δf the width of the resonance, measured at half the maximum amplitude. A very high Q means that the resonance is very sharp and the operating frequency high. Figure 2.3 provides an example for both a low and a high Q resonance. It is very obvious that excellent stable clocks must have a very high Q factor in order to provide something close to a single output frequency. For all our purposes we have a second important requirement. The oscillator must be extremely stable, since the oscillation frequency determines the scale that critically depends on the absence of any drift of the clock frequency. In practice it is a lot easier to generate an oscillator with a sufficiently high Q than to make the clock ultimately stable.

2.1.1 Clock stability

An ideal clock will have a practically infinite Q and the frequency of operation will never change. It is clear that this is impossible. Therefore, we need a criterion for how well a given clock is operating and how stable it is. A simple approach uses a larger number of successive frequency estimates and computes the variance σ^2 of these measurements, by setting

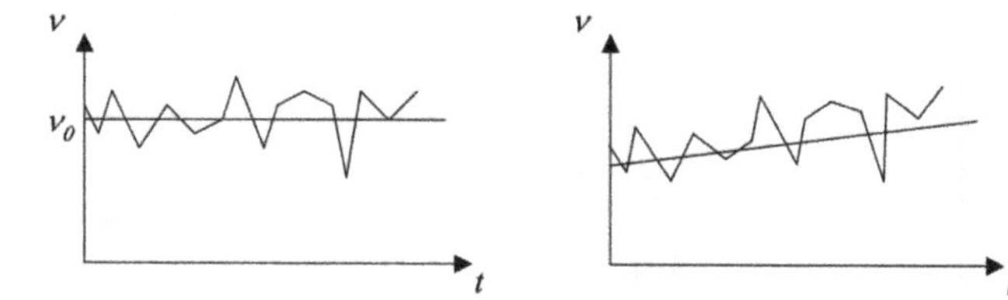

Figure 2.4. Illustration of the determination of the scatter of a stable clock (left) and a drifting clock (right), where the calculation of the variance becomes meaningless.

$$\sigma^2 = \frac{1}{n-1}\sum_{i=1}^{n}(\nu_i - \bar{\nu})^2. \tag{2.20}$$

A larger scatter of the individual measurements about the mean value $\bar{\nu}^2$ results in a larger σ^2 and hence characterizes the clock as inferior. If the clock is also drifting, it means that the definition of a mean value for the clock frequency is rather meaningless. Figure 2.4 illustrates this situation. In order to characterize the drifting clock it is very useful to compute the *Allan variance* [18], which is defined as

$$\sigma^2 = \frac{1}{2(n-1)}\sum_{i=1}^{n}(\nu_{i+1} - \nu_i)^2. \tag{2.21}$$

The Allan variance sums over the difference of neighboring measurements. Another important property is that the Allan variance is usually evaluated over intervals of time with the duration τ. When the Allan variance is calculated as a function of τ, it presents information about underlying noise processes [13]. Usually, it is useful to evaluate the square root of the Allan variance, since the Allan deviation has the same units as the measurement quantity of interest itself. Figure 2.5 illustrates this property. White phase or phase flicker noise averages down as τ^{-2}, white frequency noise as τ^{-1}, flicker frequency noise does not improve from longer averaging and frequency drift will cause the Allan deviation to diverge with longer averaging intervals proportional to τ. Usually all these processes are present in every practical oscillator. Therefore, it is important to see where the knee of the Allan deviation is located. It identifies the time interval over which the clock is most stable and lowest in noise. The computation of the Allan deviation is not limited to the characterization of clocks.

2.2 Clocks used in space geodesy

The techniques of space geodesy are all based on measurements of the time of flight of electromagnetic signals, propagating at the vacuum velocity of light c_0. We recall that c_0 is a constant and traceable to fundamental constants ($c_0 = 1/\sqrt{\mu_0\varepsilon_0}$). In order to obtain the desired distances and directions at very high resolution, one has to time either the arrival of a signal or the duration of the signal travel time. The achieved resolution depends on the bandwidth of the detector, the probe signal and the clock resolution. While this is one essential property of the measurement systems of space geodesy, it is not the full story. Further requirements are the stability of the

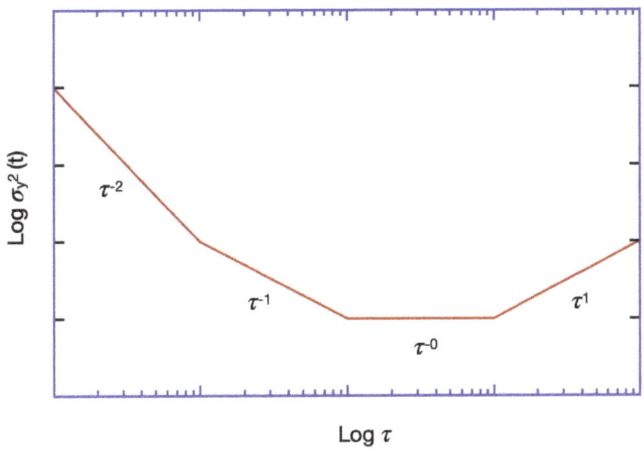

Figure 2.5. Typical noise processes and their characteristic slopes from the Allan Deviation calculation. From left to right there is white phase or flicker phase noise, white frequency noise, flicker frequency noise and random walk frequency noise.

clock and the measurement process, including the detector setup, as well as the absence of systematic errors.

In practice it turns out that it is much simpler to achieve high resolution than it is to provide the long-term stability and the absence of systematic errors. Clocks can be used in different ways. They may provide time or they may provide the frequency of the clock generator, which is then subsequently used to facilitate the physical measurement process. Time, for example, is required to track a satellite target or to synchronize a network of radio telescope for a measurement schedule. Frequency, on the other hand, is required to switch measurement gates of short duration on and off, to derive mixing frequencies for local oscillators and to derive sufficient small subdivisions of the time pips from the clock in order to obtain high- resolution range measurements.

Not all of the timing signals are required to have ultimate stability and resolution. Different tasks have different requirements and so it is not surprising that the techniques of space geodesy use a variety of different clocks for their purpose. Apart from economic considerations there are several properties of clocks to consider. These are:

- **Size and weight**: A clock used in a fixed ground laboratory can be larger and heavier than a clock that is used in a mobile or portable setup.
- **Voltages**: Clocks that for example require high voltages to be operated are certainly unsuitable for satellite missions.
- **Mean time between failures**: A clock that is operated on a limited supply of particular atoms will have a limited time of operation, which may be incompatible with the desired application.
- **Resolution**: If the clock resolution is insufficient for a particular application, it cannot be used for that purpose.

- **Stability**: Different clock concepts offer different levels of stability and that may limit the choices for a particular experiment.
- **Accuracy**: Some clock concepts provide a higher level of accuracy than others, so this has to be taken into account as well.

In the following sections we will discuss the main types of clocks that are used in space geodesy and outline their area of application.

2.2.1 Quartz clocks

Starting around the middle of the 20th century, quartz clocks came into existence. They use mechanical vibrations on a quartz mono-crystal by exciting a narrow fundamental eigenmode of oscillation. They can be considered similar to a resonant system, made from a coil (inductance L) and a capacitor (capacitance C), whose resonance frequency would be

$$f = \frac{1}{2\pi\sqrt{LC}}. \tag{2.22}$$

In order to maintain the oscillation, a transistor amplifier is required, which feeds back onto the resonant circuit and compensates for the losses of the system. A standard coil and capacitor also provide a resonance, but they do not offer a high Q-factor, which means that the resonance would be neither very sharp, nor stable in frequency. Substituting the coil and the capacitor for a monolithic SiO_2 mono-crystal establishes thickness variations via the piezo electric effect and results in resonant oscillations, which provide several advantages. First and foremost, it brings the Q-factor up from much less than 1000 to values between 20 000 and 200 000. Apart from operating on higher frequencies, the oscillation becomes much more stable, while the entire circuit is physically much smaller and much less susceptible to environmental temperature changes. These properties can be further improved by operating the quartz on a higher harmonic of the fundamental frequency and by placing the circuitry inside a small temperature-controlled compartment. Adding a buffer amplifier to the output line avoids disturbances, as it isolates changing loads from the oscillator. Today small crystal oscillators are found in about every modern wrist-watch and many electronic gadgets. Historically the advent of quartz oscillators made frequent clock adjustments unnecessary, due to the much enhanced oscillator stability.

Figure 2.6 shows a simplified circuit diagram of a quartz clock oscillator with varactor tuning. Naturally, not all quartz clocks show the same performance. In space geodesy we use carefully designed high-end oscillators with a Q-factor of about $Q \approx 10^6$. Nevertheless there are some disadvantages of quartz clocks as well. First of all, the frequency is not so well defined. The resonance depends on the physical size of the quartz and there are limitations in providing accurate dimensions. Furthermore, when inserted into the electronic circuitry some unavoidable detuning effects result. The oscillation frequency also tends to change with the aging of the crystal structure and by adsorption and desorption of atoms from the surface of the

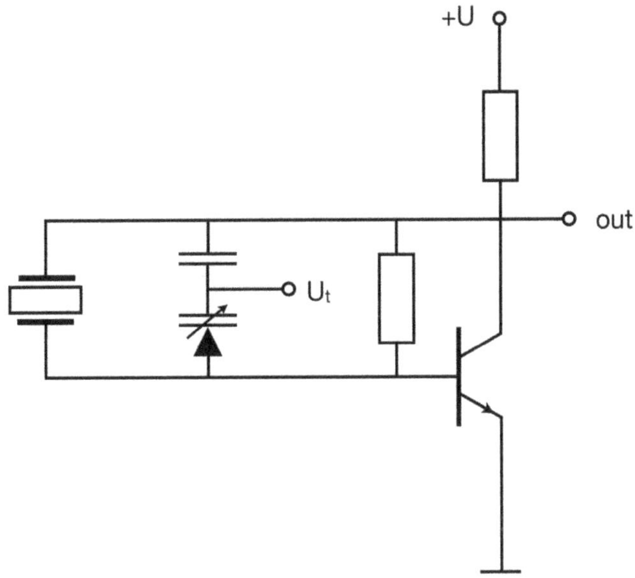

Figure 2.6. Simplified functional diagram of a quartz oscillator with varactor tuning.

quartz as time goes on. Typical values for the frequency drift of a high-end system are around

$$\frac{\Delta f}{f} < 10^{-10} \text{per day.} \tag{2.23}$$

Quartz oscillators have very good frequency stability and very clean oscillations in the very short term because of the mechanical inertia of the vibrating quartz. Therefore, they are often found as auxiliary oscillators in other clock concepts. Despite the fact that the oscillation frequency of a quartz oscillator is determined by the physical size of the quartz crystal, the oscillation frequency of the oscillator can be slightly tuned over the range of a few parts in a million. By adding a varactor diode (variable capacitance diode) to the oscillator circuit, it is possible to slightly shift the oscillation frequency proportional to a small applied DC voltage. Usually, this feature is often used in feedback systems in order to keep the resonance either in the right spot or at the maximum signal level. While quartz clocks lack a mechanism that strictly defines their operation frequency with high resolution and accuracy, they still make exquisite *clean-up* oscillators.

2.2.2 Atomic clocks

2.2.2.1 The cesium clocks

When it comes to the need of accurate clocks, we have to explore atomic clocks. Unlike what their name seems to suggest, atomic clocks do not operate on any dangerous radiation. They rather exploit the property of all atoms that they can come in higher or lower energy states and that these energy states are only changing

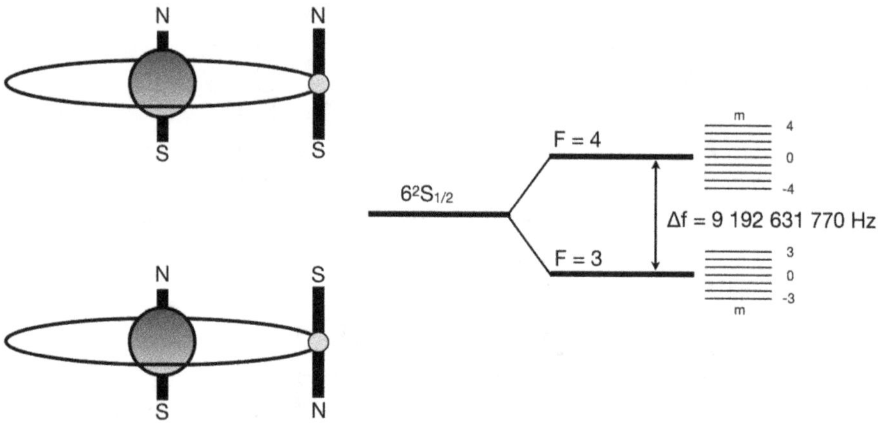

Figure 2.7. Simplified illustration of the cesium clock transition.

in very discrete and well-defined steps. That means that a suitable atom, which offers an easily accessible and exquisitely sharp transition, can act as the required frequency reference that quartz oscillators are lacking. Furthermore, by combining the properties of a quartz oscillator and a suitable atomic transition, we can finally construct a very stable and well-defined clock, which is capable of maintaining a stable time scale over a long time. The most prominent atomic transition exploited for this purpose is the hyperfine structure transition between the $F = 4$ to the $F = 3$ ground state in the ^{133}Cs *(cesium)* isotope. It corresponds to a flip in the magnetic moment of an electron from the higher energy state, where both nucleus and the valence electron have parallel magnetic moments, to the lower energy state where nucleus and the valence electron have antiparallel magnetic moments. The energy difference of this transition is in the microwave regime and is defined as 9 192 631 770 Hz. It is fair to say that this value is tailored to match as closely the length of the SI second as it was used before the days of the atomic clocks. Cesium was chosen because this transition is stable, well-defined and technologically readily accessible. Figure 2.7 illustrates this clock transition graphically. So far we have looked at the transition alone. Now we need to construct the respective clock in order to utilize this concept. In order to exploit the cesium transition in a clock setup, it is required to take *Cs* atoms from a reservoir and condition them to be in the $F = 3$ ground state. This is achieved by heating the cesium atoms in a furnace, so that a constant beam of atoms is flowing out from the reservoir. The preconditioned atoms are filtered for the ground state by utilizing an inhomogeneous magnetic field, interacting with the cesium atom dipoles. Then they are transferred to a resonant microwave cavity, where they are subjected to the radiation of a quartz oscillator, which is tuned to hit the clock transition of the cesium atoms. Depending on how well the quartz oscillator is matched to the clock transition, some or many atoms are pushed up into the upper state of the clock transition. As the atoms are moving out of the microwave cavity, they are subjected to another inhomogeneous magnetic field in order to separate the atoms in the excited clock state from the atoms still remaining in the ground state.

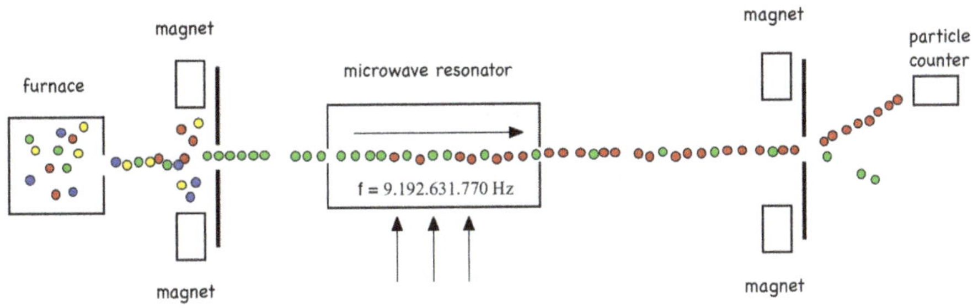

Figure 2.8. Illustration of the basic cesium clock operation. Green circles indicate the ground state, red circles the excited clock transition. Other colors represent other energy states, which are filtered out.

A detector determines the conversion efficiency into the clock transition of the cesium atoms. The detector consists of an electrode, where the cesium atoms are ionized to Cs^+. This generates a current, which is proportional to the number of incoming ions. From this established conversion rate a feedback signal is generated that adjusts the operation frequency of the quartz oscillator in order to maximize the conversion of the cesium atoms into the upper energy level. When the conversion maximum is reached, the atomic transition and the quartz oscillator operate on the same frequency, namely 9 192 631 770 Hz and the quartz oscillator can be used as the output frequency of the clock. However, we need to be aware of the fact that the quartz oscillator is continuously tuned to the cesium clock transition. Therefore, this frequency will slowly and minutely oscillate around the true clock transition frequency and the short-term stability of a cesium clock is not ideal. However, in the long-term the cesium clock will always maintain and adjust itself to hit the cesium clock transition. So a cesium clock can be considered to be long-term stable. Furthermore, only the cesium clock represents the exact duration of the SI second. Figure 2.8 illustrates the process.

While the now described process provides the principle of operation, there are few important practical considerations still necessary in order to understand the properties of a cesium clock. The ultimate clock will present itself with a very well-defined frequency. However, when we look at figure 2.8 we see a stream of moving atoms. The frequency that we observe is proportional to the energy difference between the two utilized atomic states. Moving atoms have kinetic energy. That means that the atom beam is detuning the desired atomic transition. Magnetic fields do so too, they are changing the degeneracy of the energy states. Therefore it is important to apply the inhomogeneous magnetic fields at the input and output of the resonator orthogonal to the path of motion of the atoms. Furthermore, the atoms themselves have to move as slowly as practical through the cavity. We have

$$\Delta E = h\nu_0 + E_{\text{kin}} \quad \text{and} \quad E_{\text{kin}} = \frac{m}{2}v^2. \tag{2.24}$$

Observation time or better interrogation time and spectral resolution are inversely proportional to each other and this feature is illustrated in the sketch of figure 2.9.

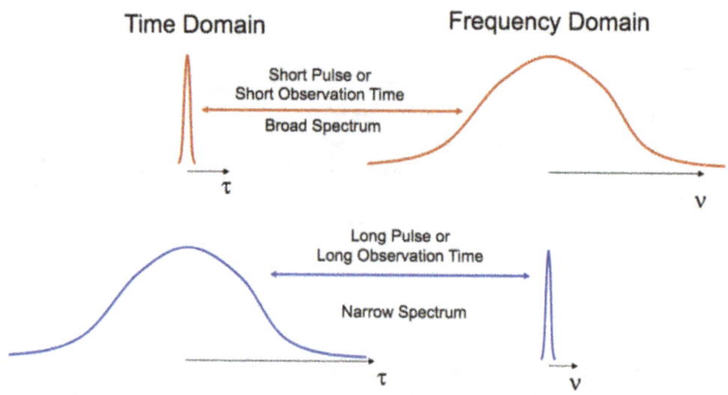

Figure 2.9. The relationship between interrogation time and spectral resolution.

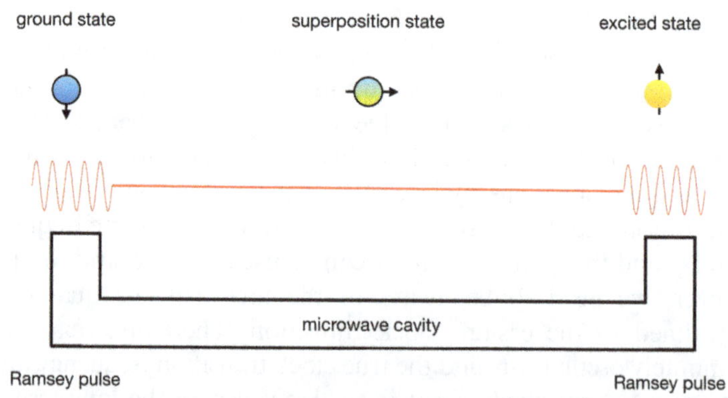

Figure 2.10. Illustration of the quantum mechanic excitation process in a cesium clock. The atom in the ground state moves in from the left. When it experiences the microwave pulse it goes into a superposition state. The second microwave pulse determines whether the atom goes into the excited state or remains in the ground state.

Processes, which are very short in the time domain cover a large spectrum in the frequency domain. Conversely, processes, which occupy a long interrogation time in the time domain are very sharply defined in the frequency domain. However, on the other hand, we cannot make the atom velocity too small, as gravity would take over and the atoms would not be able to pass through the microwave cavity. Apart from these complications, we need to ask ourselves, how narrow can the probability distribution of the excited atoms at the detector be? This also determines the clock resolution in a critical way [15]. At this point it is useful to recall that atom excitations are inherently quantum mechanical processes and this allows us to view the involved processes in an entirely different way, which is sketched in figure 2.10. The atom enters the microwave cavity from the left and is exposed to a short pulse ($\pi/2$-pulse), which transfers the atom into the superposition state, where both the spin up and spin down state are present. This can be pictured as the spin being halfway up and precessing around the 'equator' of the atom as it progresses along the cavity. When it reaches the second interrogation region, it experiences a second

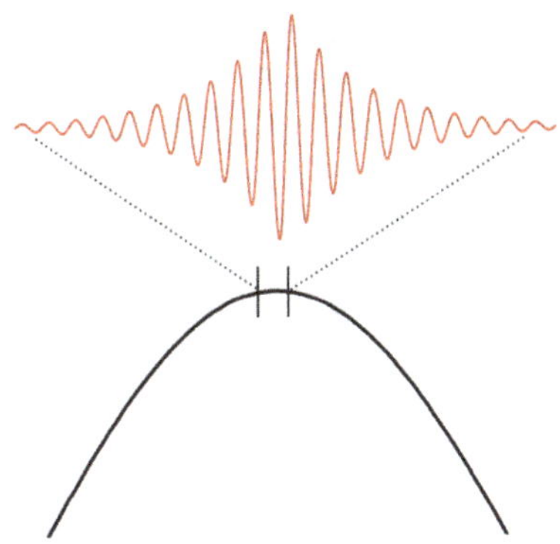

Figure 2.11. Sketch of the transition probability of a cesium atom exposed to resonant microwave excitation (black) and the same process when two Ramsey pulses are applied.

$\pi/2$-pulse, which has the same phase relationship relative to the first pulse, but zero amplitude in the transition region. When the precessing spin vector has completed an integer multiple of revolutions between the two pulses, it comes out in the excited state. Otherwise it drops back to the ground state. The result is a much sharper discrimination of the match between the atomic state and the interrogating microwave frequency and gives rise to an interference pattern. The point of operation of the feedback system is then set to the middle of the slope of the largest excursion. This is illustrated by the transition probability as shown in figure 2.11. It is now obvious that the successful transition to the upper state depends critically on both, the propagation velocity of the atom and the frequency of the Ramsey signal. A standard commercial cesium clock like the HP5071A reaches a relative uncertainty of two parts in 10^{13} after about one day. In order to improve this situation, one has to reduce the velocity of the atoms along the interrogation cavity even further. In order to achieve this, the thermal energy of the atoms have to be controlled in a much better fashion and the interrogation zone has to be set up vertically. This concept is realized in the cesium fountain clocks. Although very similar in the basic physical operation principle of a cesium clock, the design and operation of a cesium fountain clock is vastly different. The central part of a cesium fountain clock is a magneto-optic trap (MOT), where a cloud of cesium atoms is trapped and laser cooled. For an explanation of laser cooling see section 2.2.3. The laser cooling reduces the kinetic energy of the entire ensemble of atoms and reaches a temperature of values near 10 μK. Once sufficient cooling is achieved, the MOT is turned off and the atoms are gently pushed upwards by laser interaction. On the way up they pass through the microwave cavity and are subjected to the first $\pi/2$-pulse. After about 1 m of upwards motion, the cloud of atoms comes to a halt, before gravity causes it to reverse the direction and drop downwards, passing through the microwave cavity a second time and being subjected to the second $\pi/2$-pulse. The total interaction time lasts for about 1 s. When the cloud has passed the microwave cavity it reaches the interrogation region, where laser beams are stimulating fluorescence in each of the two energy states in two subsequent measurements. From the intensity ratio, the conversion efficiency is then established, before the process is started all over again. Since the atom temperature of the cooled atoms is homogeneous and in the μK regime, the velocity of the atoms can be well below a m s^{-1}. That allows the cesium fountain clocks to exhibit an instability of the order of 1 part in 10^{16} after one day, about a factor of 100 more stable than the room temperature cesium clocks, discussed above [8].

Cesium fountain clocks are the most stable primary frequency standards and they represent the best continuous realization of the SI second. Primary standards have a fully characterized error budget. They are usually operated by the National Metrology Institutes (NMI) and provide the scale of the SI second.

2.2.2.2 The rubidium clocks

An alternative concept is realized by utilizing a suitable hyperfine transition in the atom species rubidium (Rb). A gas discharge of ^{87}Rb in a vapor cell provides a light source at a wavelength of 780 nm. This emitted light is on the red end of the visible spectrum. Another vapor cell, filled with the ^{85}Rb isotope acts as a filter and removes radiation from the light source, which is not in the 780 nm regime. The remaining light excites another set of ^{87}Rb atoms in a vapor cell from the ground state into the $^5P_{3/2}$ state. This state has a short lifetime and the cell is located inside a microwave cavity. The electrons quickly relax into the $F = 2$ upper hyperfine state of the $^5S_{1/2}$ clock transition. A photodetector, which is located behind the resonator, sees changing light levels of the 780 nm radiation. As the rubidium atoms are pumped up into the ^{5}P state, the absorbed photons are missing from the detector, which measures a lower voltage. If we assume that the clock transition of all available rubidium atoms is populated, no light is absorbed and consequently the detector records a higher voltage. In analogy to the cesium clock a rubidium clock comes with a feedback mechanism, which fixes the resonance frequency. A microwave oscillator signal, tuned to the clock transition of 6.834 GHz, is injected into the microwave cavity containing the rubidium cell. When the injected frequency matches the clock transition, it stimulates the emission of the excited atoms back to the ground state. As a result the initial pumping process kicks in and the light on the photodetector reduces again. An additional buffer gas composed of noble gases inside the test cell, reduces the atom velocity and therefore ensures a long interaction time. A stable set-point at the desired resonance frequency is reached, when the feedback mechanism keeps the light intensity low. Figure 2.12 illustrates the functional principle of a rubidium clock schematically.

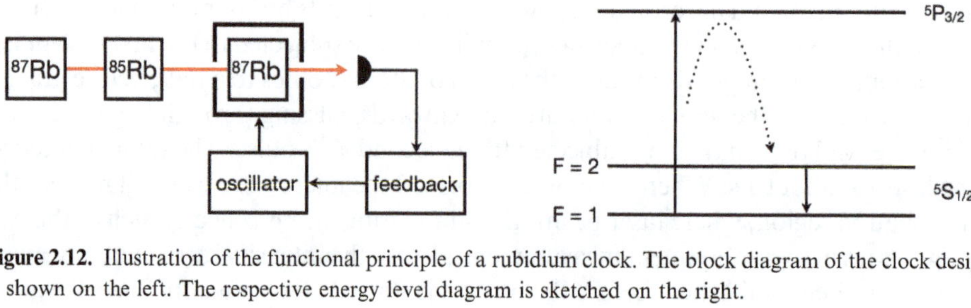

Figure 2.12. Illustration of the functional principle of a rubidium clock. The block diagram of the clock design is shown on the left. The respective energy level diagram is sketched on the right.

Compared to the cesium clock, a rubidium clock does not exhibit the same long-term stability and frequency resolution. Nevertheless, it makes an excellent small, light weight and compact frequency standard with high reliability. The short-term stability of the Rubidium clock is comparable to a commercial cesium standard, since both clock types have a feedback loop that minutely oscillates about the respective laser transition in order to provide a known frequency reference. Rubidium frequency standards are highly desirable for satellite missions. Unlike cesium clocks they do not lose their supply of interrogation atoms when they are operated. Therefore, operation times in excess of 10 years are not uncommon for a rubidium standard. For applications that require stable frequencies, a rubidium standard is between one and two orders of magnitude lower in stability, when compared to a commercial cesium standard, such as the HP5071A. In terms of achievable accuracy with respect to the actual oscillation frequency, a rubidium clock provides values of $\Delta f / f \approx 10^{-9}$. For many applications in space-borne satellites this is sufficient. Rubidium clocks can be operated both as an active or a passive system[2]. In practice however, only the passive system, as outlined above, is technically relevant.

2.2.2.3 The hydrogen maser

Clocks are used for very different purposes, not only to record or represent the passage of time. For some applications a clock provides a constant and well known frequency, for other applications a clock provides markers for specific instances or as it is put in space geodesy, exact markers for a precisely specified epoch. A range measurement to a satellite for example is only valid for a specific epoch, pinpointing the position of the satellite in its orbit. The number of clock oscillations however, is related to the actual range measurement. Radio interferometry, determines the time delay of a microwave signal between several widely spaced detectors. These measurements are carried out at microwave frequencies of 8–15 GHz, but the evaluation of the time delay, for practical purposes, has to be done at much lower frequencies. Therefore, it is necessary to down-convert the microwaves and this requires frequency mixing [17]. In this context it is important to utilize a reference frequency, which is much more stable and significantly lower in noise than the detected signal of interest. The system of choice for these applications is the hydrogen maser. The actual output frequency is not so well-defined because the transition at nominally at $f = 1.420405\ 75$ GHz is not so sharp. A hydrogen maser has excellent stability, defined by the resonator and very low-noise levels, even in the short term. This makes it the only important active frequency standard in space geodesy. Figure 2.13 provides a basic sketch of the hydrogen maser design. A reservoir of hydrogen molecules (H_2) utilizes a high-frequency discharge to dissociate the hydrogen molecules and to populate the hyperfine structure transition ($F = 1$). This results in an atom beam, which is directed toward a storage bulb. The atom beam passes an inhomogeneous magnetic field. Due to the dipole moment of the atoms, only those in the proper upper state enter the storage bulb, which in turn is part

[2] What is the meaning of active and passive in this context? A passive system interrogates a quantum state, while an active system excites radiation via stimulated emission.

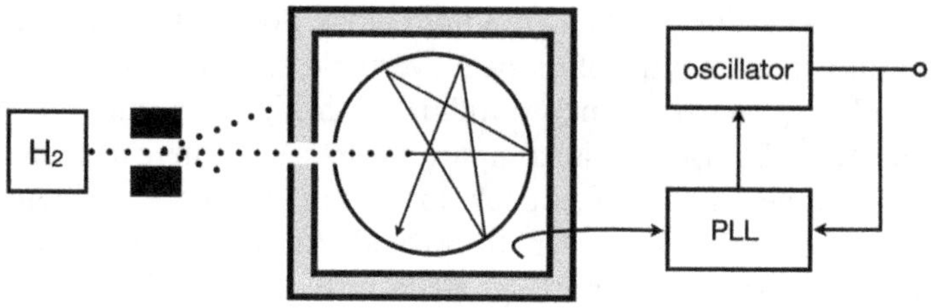

Figure 2.13. Operation principle of an active hydrogen maser. The hydrogen molecules are dissociated in a high-frequency discharge and a beam of H^+ atoms is ejected toward a storage bulb inside a microwave cavity. An inhomogeneous magnetic field filters unwanted energy states out. Inside the bulb, atoms de-excite through spontaneous emission. This causes subsequent stimulated emission from other excited atoms. The signal is picked off and a PLL synchronizes a quartz oscillator to the emission frequency.

of a microwave resonator, tuned to $f = 1.420\,405\,75$GHz. This frequency is well known to astronomers because this 21 cm wavelength is one of the most important observation signals in radio astronomy. Hydrogen is a fundamental element in the Universe. The storage bulb has a volume of about 2 l and is coated with Teflon in order to avoid recombination of the hydrogen atoms and to make the lifetime of the upper energy state and therefore the interaction time as long as possible. The longer the lifetime of the energy state of the atom in the storage bulb, the smaller is the linewidth of the transition. This is a consequence of the uncertainty relationship: $\Delta f \cdot \Delta t \leqslant 1$. A typical entry velocity of the atoms into the storage bulb is 2 km s^{-1} and the lifetime of the upper state is between 0.5 and 1 s. Due to spontaneous emission we obtain a weak microwave radiation, which is in resonance with the microwave cavity around the storage bulb and causes stimulated emission to the other excited atoms in the storage bulb and therefore coherent amplification of the microwave radiation. The rate of influx of new excited atoms and the rate of removal of de-excited atoms from the storage bulb by a vacuum ion pump determines when the amplification becomes stationary. Most of the microwave field is required to maintain the maser oscillation, only a very small portion is picked up by an antenna and can be used in a phase-locked loop amplifier to synchronize an external quartz oscillator to the maser frequency. The absence of any frequency tuning in this setup provides the excellent short-term stability and the large number of contributing atoms provide the low-noise level of the maser.

Figure 2.14 summarizes the properties of the clock and frequency standard realizations usually used in space geodesy in a simple diagram. The cesium clock is actively steered and it represents the definition of the SI second (B). In the short term we do not get a good stability from a cesium clock, but in the long-term such a clock provides excellent stability. A quartz oscillator provides low noise, but its frequency of oscillation is not well-defined and it drifts considerably (D). However, when locked to a stable frequency standard, they make excellent frequency generators. Hydrogen masers (A and C) exhibit low noise already in the short term, but their frequency of operation is not so well-defined. This is caused by the mechanical properties of the microwave resonator, which encloses the storage bulb. Subtle differences in their dimensions, variations of temperature and insufficient shielding

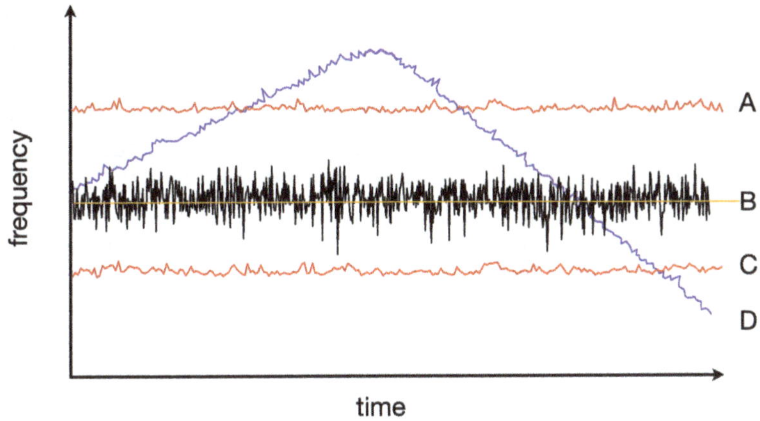

Figure 2.14. Schematic illustration of clock properties. A maser is stable and has low noise (A). However, when it is switched off and back on, it may run on a slightly different frequency because of the width of the transition (C). A free-running quartz clock is very low on noise, but the frequency is not defined (D). A cesium clock is constantly steered to the cesium transition, which defines the length of the SI second (B). It is long-term stable, but noisy in the short-term.

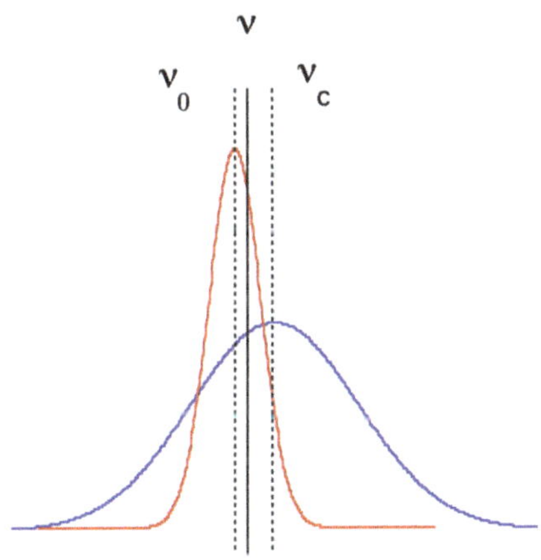

Figure 2.15. Schematic illustration of the clock detuning effect. The atomic transition and the cavity resonator of an atomic clock are detuning each other, depending of the quality factor of each resonance. The red curve stands for the atomic transition with the center frequency ν_0. The blue curve is the cavity resonance ν_c, which has a much lower Q. The actual oscillation frequency is ν and it is pulled a little away from ν_0.

from magnetic fields cause a small detuning of the resonance frequency as well as the Q-factor of the cavity. There are limitations in fixing these environmental properties to the tight requirements of an accurate clock. At this point it is important to take a moment and consider the interaction between a clock transition, the involved cavity resonator and the actual operating frequency of a clock. Since these properties are very general, they apply to all clock realizations and is illustrated in figure 2.15. The clock transition ν_0 has a high Q, which means that the resonance is sharp, similar to the red curve in figure 2.15, which exhibits a small spread and a well-defined high peak. The cavity resonance ν_c, on the other hand, is much broader and the peak is considerably less defined. Since both peak values are not on the same spot on the horizontal frequency axis, the actually observed clock frequency is somewhere in between the two resonances ν_0 and ν_c. The resonance with the higher Q pulls the oscillation frequency more than the resonance with the lower Q. In our illustration we can see that the actual oscillation frequency ν is closer to the maximum of the red curve. Conversely, this means that a resonator with a very low Q causes much less detuning. However, on the other hand, if the cavity Q in an active system like a hydrogen maser is too low, it will not be possible to maintain the stimulated emission. This combination of atomic transition and resonant cavity can be considered a coupled oscillator. The effective detuning can be calculated by

$$\Delta\nu = \nu - \nu_0 = \frac{Q_c}{Q}(\nu - \nu_c). \tag{2.25}$$

We can conclude that a nearly unperturbed transition frequency is only obtainable in a passive system, when the Q of the cavity is almost vanishing. In active systems, such as a laser and also a maser, the oscillation frequency is strongly depending on the cavity. Lasers and masers are therefore no good frequency references, although their linewidth can become very narrow. However, when they are locked to a suitable reference, they can become excellent oscillators. For the frequency detuning of a passive system we can alternatively write

$$\Delta\nu = \nu - \nu_0 = \left(\frac{Q_c}{Q}\right)^2(\nu - \nu_c)\left(2\frac{\Delta P}{P_0}\right). \tag{2.26}$$

This expression is very similar to equation (2.25). It was extended by a factor that accounts for the fluctuation of the signal power in the resonator. In summary, an active system, operating on stimulated emission, requires a low loss resonator of high Q. This results in a stationary oscillation of a sharply defined frequency. The passive system on the contrary obtains the reference to a defined frequency by the interrogation of a particular atomic state. This can be facilitated by

- absorption;
- re-emission;
- test for a particular particle state;
- indirect test; for example, the observation of changes in brightness of test radiation (e.g. the rubidium clock).

Where is the limit for the stability of active or passive oscillator? Active oscillators are limited by thermal noise of the resonator. A statistical expectation value for the frequency fluctuations is

$$\langle\sigma^2\rangle \approx \frac{1}{Q^2}\frac{k_B T}{2P_T} \quad \text{for} \quad h\nu \ll k_B T, \tag{2.27}$$

where Q is the quality factor of the atomic transition, namely $(f/\Delta f)$, P the power of the stimulated emission radiation, τ the interaction time interval, T the temperature and k_B the Boltzmann[3] constant. The statistical expectation value for the frequency fluctuations of a passive resonator can be formulated in a similar manner. Such a system is limited in stability by the shot noise (power) of the auxiliary oscillator.

$$\langle \sigma^2 \rangle \approx \frac{1}{4Q^2} \frac{h\nu}{P_T} \quad \text{for} \quad h\nu \ll k_B T \tag{2.28}$$

The power is given by $P = nh\nu$, that is, a number of n transitions per unit time stimulated by the external oscillator. When we compare the expectation values for the frequency uncertainty between equations (2.27) and (2.28) under comparable situations, we find that theoretically the noise of a passive system can be lower. However, that does not necessarily mean that a robust working system can technically be realized.

2.2.3 Optical clocks

Why would we care about optical clocks? In order to improve the stability of a clock as well as the resolution, it is very obvious that higher frequencies of the clock oscillator provide a great advantage. On the contrary, it would be impossible to obtain a resolution of 1 ms from a grandfather's clock, which operates on a pendulum oscillator with a frequency of 0.5 Hz. The figure of merit is reached, when the quantity $\Delta f/f$ is minimized. The best way to achieve that is to make the frequency f of the clock oscillator large. This is essentially what we have already done in the previous sections, when we went from a few MHz of a quartz clock to the GHz frequencies of the atomic clocks. By going up into the optical domain, we increase the oscillator frequency by roughly another 4–5 orders of magnitude. The design and operation of an optical clock is a very wide field and in this book we limit ourselves to the basic design and operation concept of a single ion ^{171}Ytterbium clock. A comprehensive review of optical clocks can be found in [12]. Apart from the advantage of a higher operation frequency, optical clocks are cooling the ions or atoms to extremely low temperatures, so that the detuning from the kinetic energy of the probe species becomes negligible. Another typical difficulty of optical clocks, namely the exact determination of the actual oscillation frequency has been removed with the advent of optical frequency combs [9]. Schematically we can identify three main functional components of an optical clock. This is illustrated in figure 2.16. The first element of an ytterbium ion clock is a Paul trap in which a very small number of ions are held inside an ultrahigh vacuum recipient. The Paul trap consists of a ring and one electrode at the top and another electrode at the bottom. A high-frequency generator generates an alternating electro-magnetic quadrupole field between the ring and the electrodes, which vanishes at the center of the trap and

[3] The Boltzmann constant k or k_B is a fundamental physical constant and corresponds to the quotient of the universal gas constant (R_m) and the Avogadro constant (N_A). It has the unit of (J K^{-1}) (energy per Kelvin) and a value of $k = 1.380\ 648\ 8(13) \times 10^{-23}$ J K^{-1}.

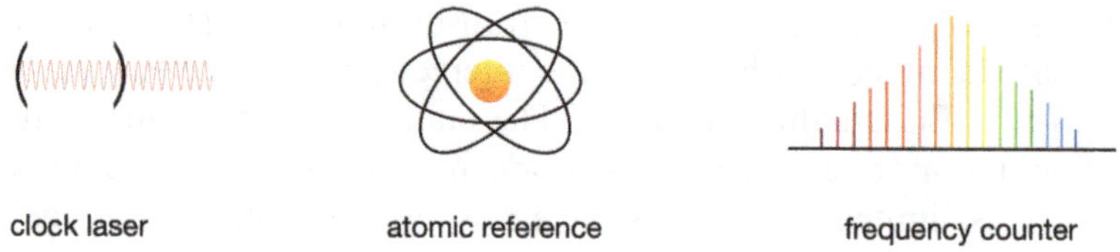

Figure 2.16. Basic setup of an optical clock. The atom or ion provides a stable narrow transition, which is interrogated by the tunable clock laser. When the clock laser is in resonance with the optical transition, the frequency comb down converts the optical clock frequency into the microwave regime.

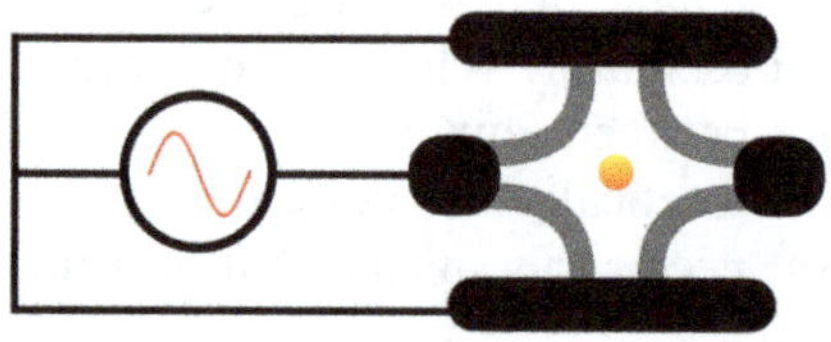

Figure 2.17. Schematic cross-section of a Paul trap, which has a vanishing electromagnetic quadrupole field at the center. Therefore, it can hold the Ytterbium ions in nearly fixed positions for a very long time.

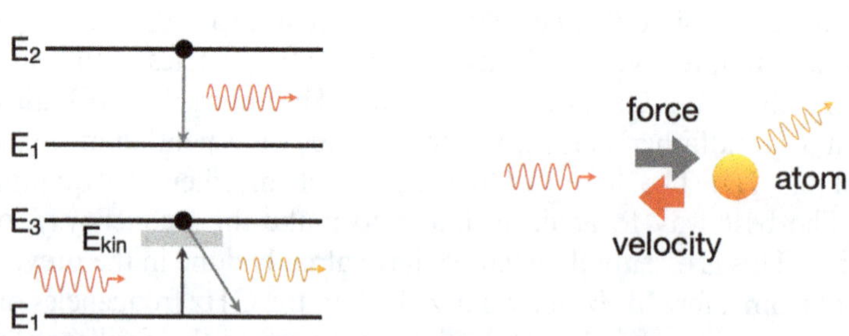

Figure 2.18. Energy level diagram illustration of the laser cooling process. It requires some kinetic energy to enable the excitation of an atom to a slightly detuned higher state (left). An alternative explanation would be that the radiation pressure from the cooling laser slows the atom down. The atom releases the acquired energy quickly by emitting a photon of higher frequency in an arbitrary direction.

holds the ions in nearly fixed positions for a very long time. Figure 2.17 illustrates this concept. Once the ions are trapped, they can be conditioned for the next step, which is the laser cooling, in order to reduce their kinetic energy. This results in an unperturbed state for the interrogation of the chosen clock transition. Since the ions can move in three spatial directions, a total of six laser beams with the ion at the center are required. The lasers need to be tuned precisely in frequency. At the end of this cooling process, the thermal energy of the ions is reduced significantly, so that the ions are confined within a volume of about 1 mm in cross-section and their residual motion is only within 0.1μm. This is less than the wavelength of visible light. The electrical field in the center of the trap is zero, hence there is no Doppler shift on the atoms. Only a small effect from possible variations in the magnetic field are still remaining.

The laser cooling process needs some further explanations and it is illustrated in figure 2.18. When an atom transits from an excited state to a lower energy state, it emits a wave packet corresponding to the energy difference. For a hyperfine transition as for the atomic clocks, the energy difference is small and the radiation is in the microwave regime. However, when the energy difference is higher, the emitted photon may be in the regime of the visible light. This is sketched in the figure 2.18 on the top left. Conversely, when a photon is in resonance with a transition, it is absorbed and the atom changes to an excited state. The process of laser cooling exploits this property. A cooling laser is tuned to a suitable transition and the energy of the pumping photons is deliberately set low. In order to transit into an excited state, some part of the kinetic energy of the atom in motion is required. The excited state is short-lived and the atom releases a photon of higher energy in an arbitrary direction soon after the upward transition. This is shown schematically in the lower-left side of figure 2.18. The right side of the figure illustrates the same process from a different viewpoint. The radiation pressure from an incoming photon from the cooling laser decelerates the moving atom as it reduces its kinetic energy by exerting a force on the atom. More details can be found in [2] and as a result of this process temperatures well below 1 K are achieved. In other words, the trapped ion has practically no kinetic energy left and that means that there is no detuning of the clock transition by the kinetic energy of the atom degrading the measurements.

Once the ^{171}Ytterbium ion for the clock operation is trapped and cooled, it is ready for the interrogation of the clock state. The transition itself is at 467 nm. Quantum mechanically it is a weakly allowed transition and only 2.4 Hz wide. The measurement process consists of a repeatedly carried out test sequence, between a cooling laser and the interrogation laser. Figure 2.19 illustrates the procedure in a block diagram. The cooling transition and the clock transition share the same lower energy state. Therefore, the electron can either be pumped up into the cooling state or into the clock state. The lifetime of the cooling state is short and the returning electron causes a fluorescence signal, when it returns. This means that the success of a cooling cycle can be identified. The clock transition, in contrast, has a long lifetime, which means that the electron is missing from a number of cooling cycles

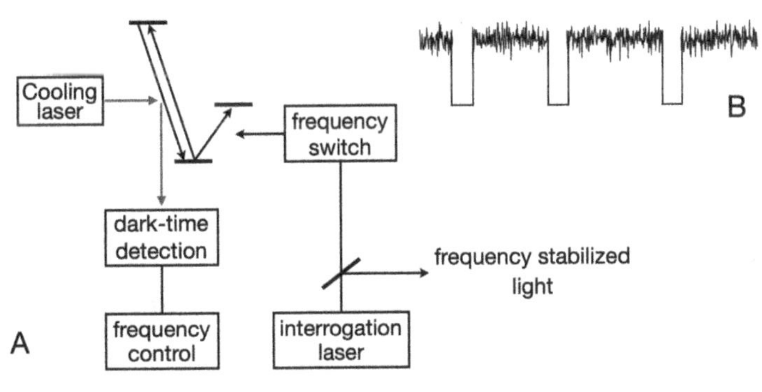

Figure 2.19. Illustration of the ytterbium clock interrogation process. A cooling laser and an interrogation laser periodically are probing two atomic states, which share the same lower energy state (A). When the electron is missing in the cooling cycle, the interrogation laser has hit the clock transition (B). Due to the longer lifetime of the clock state it takes some time for the electron to return to the ground state.

subsequent to the excited clock transition. From this fact it can be deduced that the interrogation laser has hit the clock transition so that the pumping light beam has exactly the frequency of the clock transition. So a part of the laser signal can be taken out in order to establish the actual instantaneous laser frequency from the clock laser. The process is repeated over and over again, so that the clock laser always operates on or at least very close to the respective clock transition. With the width of 2.4 Hz, the lifetime of the upper state is given as $\tau \approx 1/\Delta\nu$, which corresponds to about 0.5 s. From this operation principle we can draw a number of conclusions. An optical clock does not provide a continuous oscillation in the way the maser concept does. Furthermore, an optical clock is designed as a reference and not intended for continuous operation. The frequency estimation process is also not really straightforward and quite an involved process to facilitate. The transition is in the optical domain, but frequency counting requires the microwave domain. So there is a huge spectral range to cover. While cascades of different lasers were used in the past for this frequency estimation process, we now have optical frequency combs for this purpose.

The basis of an optical frequency comb is a femto-second mode-locked laser. With a pulse width of less than 100 fs, these lasers can have a spectral width of more than one octave. Mode-locked lasers have excellent low-noise properties in the microwave regime, where they are modulated as well as in the optical regime, where the repetition of the pulses is given by the modulation frequency as well. The optical output signal of an optical frequency comb is composed of many Fourier frequency components of the form $y(t) = A_0 \sin(\omega_i t + \varphi)$. Every of these frequency components, which feature a clean periodic signal in the time domain, show up as a very narrow pulse in the frequency domain. This is a very general property and we have used this before with the atomic clocks. Figure 2.20 illustrates this graphically. Time is displayed on the horizontal axis and the amplitude of the frequency component on the vertical axis. For the purpose of this illustration there are no units required. The important property is that the frequencies are superimposed coherently with a fixed phase relationship, which is the essence of what phase-locked actually means. When the phase relationship is arbitrary, the pattern of the resultant signal is also very irregular. This is shown in figure 2.21 with a superposition of five different frequency components. Phase locking in the laser resonator is achieved by a modulation of the cavity quality factor by an acousto-optic modulator (AOM), provided that the gain medium of the laser offers a sufficiently broad transition. An example for such a gain medium is an erbium–ytterbium-doped solid-state crystal, which operates in the infrared. It is important that more than one octave of frequency spread can be achieved. While this means that the measurements can only be taken in a comparatively small frequency window, an optical frequency comb, nevertheless can estimate the optical frequency precisely. The microwave frequency at which the laser cavity is modulated causes all the frequencies to have this exact spacing in the frequency domain. In the time domain this generates equidistant ultrashort pulses of much less than 100 fs width. Hence this is commonly referred to as the *repetition rate* of the comb. The time domain and frequency domain representation of an optical frequency comb are Fourier pairs. Figure 2.22 illustrates this property. The visible

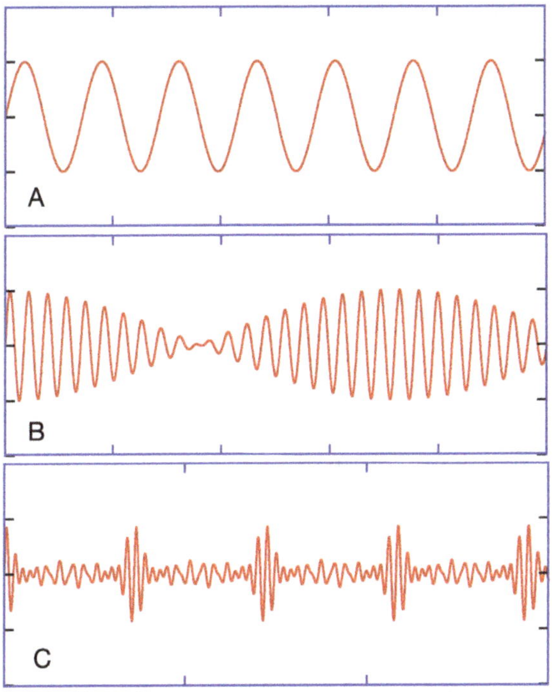

Figure 2.20. Illustration of the frequency locking process. A single harmonic sine wave has a continuous power level in the time domain (A). When two frequency components are present, we observe a beat note (B). When seven frequency components with a fixed phase relationship are composed together, we already observe the beginning of a repeating pulse pattern (C). In an optical frequency comb, hundreds of modes are mixed together.

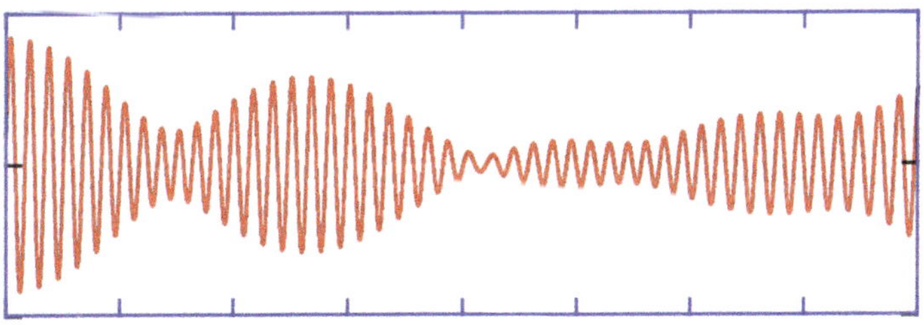

Figure 2.21. Example of a superposition of five different frequencies with arbitrary phase relationship. The resultant pattern does not produce a unique pulse pattern.

part of the frequency comb spectrum is indicated by the gray columns. The frequency spacing ω_{rep} continues all the way down toward the zero point, although there is no detectable signal in the regime between the visible part and the origin of the frequency scale. At the very low end, there is usually an offset between this repetition frequency and the origin of the frequency scale, known as the *offset frequency* ω_0. The repetition rate and the offset frequency are the only two quantities necessary to characterize an optical frequency comb. With $\omega_{\text{rep}} = 2\pi f_{\text{rep}}$ for the offset frequency we get

$$\frac{\omega_o}{2\pi} = \frac{\Delta\varphi \mod 2\pi}{2\pi} f_{\text{rep}}. \tag{2.29}$$

And the frequency f_j of the component j in the comb then is

$$f_j = f_0 + j \cdot f_{\text{rep}}, \tag{2.30}$$

where j is the number of teeth in the comb up to the desired frequency. How is such a comb operated in the laboratory? In order to measure an optical frequency, the comb and the frequency of interest must be covered under the gain curve of the comb laser, that is, they both need to be literally inside the regime of the gray columns in figure 2.22. The teeth of the comb above and below the probe signal of interest are beating with the unknown frequency, which is smaller than the repetition rate of the comb. For a typical repetition rate of commercial combs of say 200 MHz, one would obtain two frequency values of $0 \leqslant f \leqslant 200$ MHz. By filtering out one of them, the beat note of interest is easily accessible with common frequency counters. The application of equation (2.30) yields the desired frequency of the unknown optical signal. The offset frequency f_0 is directly available from the comb, while the ambiguities for j can be removed by making several measurements, each with a slightly altered repetition rate.

The determination of the offset frequency f_0 utilizes an interferometric setup, usually known as an *F–2F interferometer,* and this is an integral part of an optical frequency comb. The basic concept is a simple optical setup as shown in a functional sketch in figure 2.23. To produce a valid offset frequency the comb needs to cover at least one octave. Some of the pulsed light is taken out and passed through a frequency-doubler crystal. This shifts the frequency components from the 'red' end of the spectrum toward the 'blue' end. The up-converted signal is then interfered

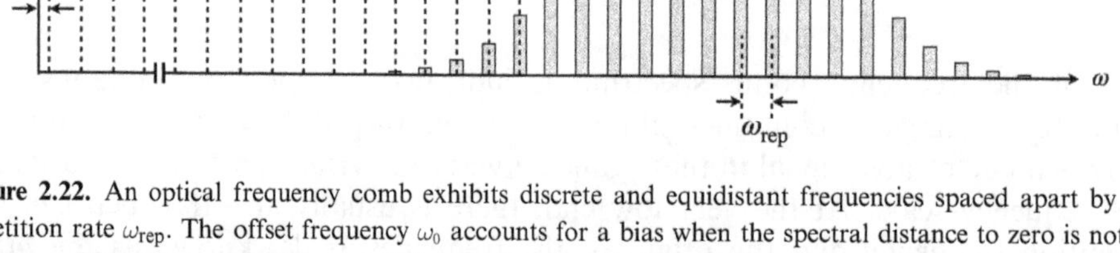

Figure 2.22. An optical frequency comb exhibits discrete and equidistant frequencies spaced apart by the repetition rate ω_{rep}. The offset frequency ω_0 accounts for a bias when the spectral distance to zero is not an integer multiple of the repetition rate. The comb pattern is only visible under the gain curve of the laser medium and needs to be at least one octave wide.

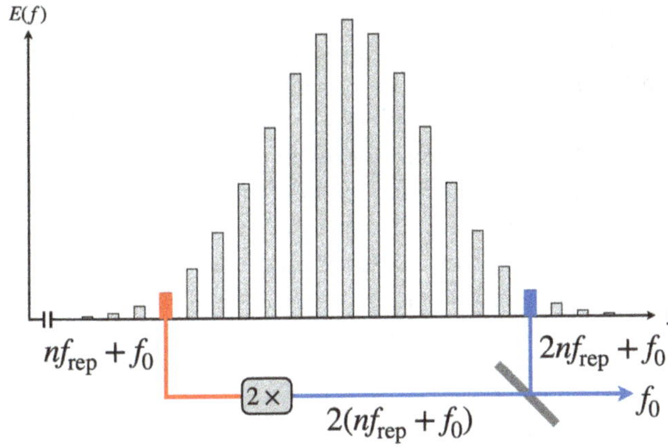

Figure 2.23. The functional diagram of an F–$2F$ Interferometer. The comb teeth used for this process need to be apart by one octave in order to produce the correct offset frequency.

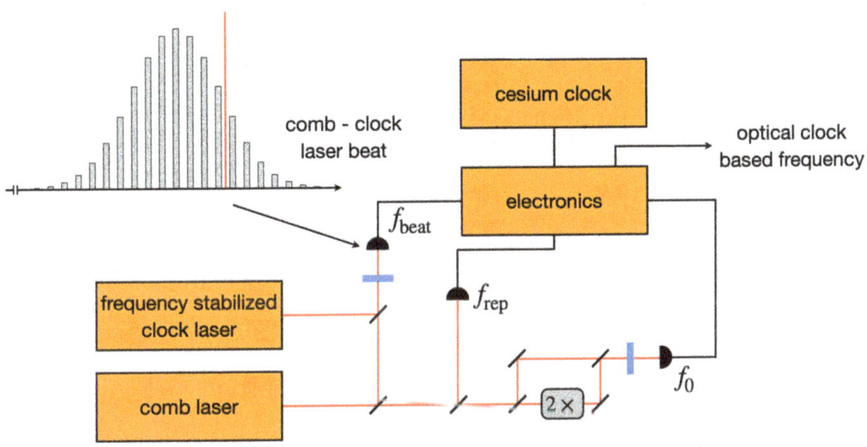

Figure 2.24. Sketch of the building blocks of a complete optical clock. While the basic concept is very straightforward and basic, the technical realization is extremely complex and bulky.

with the comb light, where the offset frequency shows up as a low-frequency component in the beat signal, somewhat below the repetition rate frequency. While the frequency doubling process doubles both, the comb teeth and the offset signal, the high-frequency end of the comb contains twice the frequency from the low-frequency end of the comb, but the offset frequency is of course not doubled at this end.

With all the building blocks for an optical clock introduced, we can now construct a full optical clock. This is outlined in figure 2.24. From the left side we have the input of the frequency-stabilized clock laser and the optical frequency comb. The beat signal between the clock laser and the comb provides the offset between the unknown clock laser frequency and a tooth of the frequency comb. Another part of the frequency comb pulse sequence goes to a photodetector and the repetition rate is established. The final part of the comb pulse sequence passes through the F–$2F$ interferometer and the offset frequency is determined. All these functions require a significant amount of control electronics and locking feedback loops. Furthermore, there is also the need for an external microwave frequency standard, like a cesium clock or a hydrogen maser. The final output product is an electronic frequency output, which can be directly traced back to the optical atom transition. It is evident that an optical clock does not run continuously like its microwave counterparts. It always needs a 'flywheel' oscillator to produce a high-quality reference. In order to provide an impression of the required hardware setup of an optical clock, a partial view of an installation at the University of Singapore is provided in figure 2.25. The figure only shows a part of the optical table. The various cooling lasers themselves are on another table and the light is transferred over in a bundle of optical fibers. The required electronics fills at least two electronics racks, each of them 2 m high and the frequency comb is yet on another optical table.

The realization of optical clocks in general is a great technical achievement since they make time the outstanding SI unit with the highest resolution of all. One of the technical consequences then is that time-of-flight measurements and frequency measurements are high-resolution measurements too. In section 2.5 we will look at an application in more detail, where the comparison of optical clocks allows for the measurement of a tiny relativistic effect. Figure 2.26 illustrates the development of the achieved accuracy of atomic clocks over the last 70 years. The curve in black roughly marked the progress of the microwave clocks, which technically level out at an accuracy of one part in 10^{16}, provided by the cesium fountain clock technology (black curve). Optical clocks became feasible once the process of laser cooling was established in the early eighties. Since about 2010 optical clocks have been outperforming microwave clocks. At the time of writing an accuracy around three parts in 10^{19} has been already achieved.

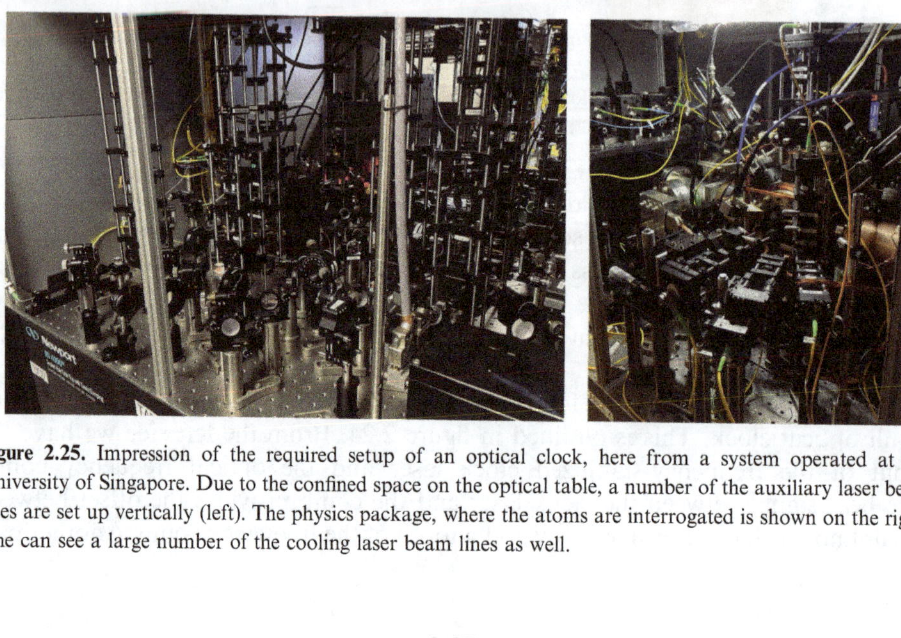

Figure 2.25. Impression of the required setup of an optical clock, here from a system operated at the University of Singapore. Due to the confined space on the optical table, a number of the auxiliary laser beam lines are set up vertically (left). The physics package, where the atoms are interrogated is shown on the right. One can see a large number of the cooling laser beam lines as well.

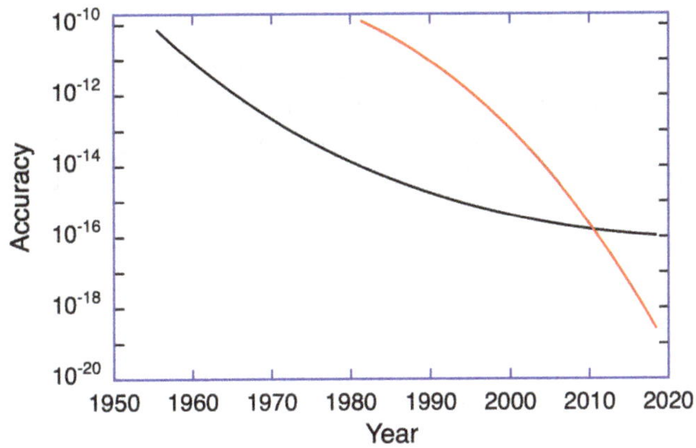

Figure 2.26. The development of atomic clocks over the last 70 years. Microwave clocks (black) have leveled out at $\delta f/f = 10^{-16}$, with cesium fountain clocks marking the top end in that class. The development of optical clocks started when laser cooling was developed (red). Optical clocks have been outperforming microwave clocks since about 2010. They have already demonstrated a level of accuracy of $\delta f/f = 10^{-19}$.

Now, let us look back at these various clock concepts. Why are they important for space geodesy and what do we really require? A typical clock drift in space geodesy is of the order of 1 ns day^{-1}. That means we lose or gain about one oscillation per day on a hydrogen maser, or the same thing expressed differently, one part in 10^{14}. At this drift rate it would take 10^9 days (more than 2.7 million years) until our clock is out by one second. Such numbers are often quoted in the general media to impress the people, when atomic clocks are the topic of interest. No clock would be able to run that long and furthermore it would not matter at all for space geodesy. What we usually use is frequency, not time. From the frequency, provided by the maser, we can establish low-noise beat frequencies for the downconversion of microwave frequencies from quasars from the range of 8–15 GHz to several hundred MHz. Here it is important that the mixer does not add noise to the faint radio signal, so that the signal to noise ratio does not degrade the correlation process. In laser ranging, the highly stable clock rate provides us with a sufficiently accurate timer gating, in order to achieve time-of-flight measurements, which are rather limited by the laser pulse width than the clock. Is time then important at all in space geodesy? In laser ranging we require time to point the telescope at the satellite for the measurement and to time-tag each single measurement. Since the satellite moves at a velocity of several km s^{-1} the requirement is moderate. A value of 1 μs for the accuracy of time is certainly enough. The *Very Long Baseline Interferometry* establishes propagation delays of quasar radiation, commonly observed by radio telescopes on different continents. For that the exact time offset between the observing stations is required. This, however, cannot be provided technically at intercontinental distances. Therefore, clock offsets are numerically solved for by adding more independent observations to the schedule. GNSS observations have the luxury of a redundant number of satellites in view. From this a clock correction for the ground segment clock can be computed.

2.3 From frequency to time

In the previous sections we have looked at different techniques to build a stable frequency reference. Different requirements led to different instrumental realizations with a variable level of complexity. While these clocks operate on very different atomic transitions and concepts, the final output of such a clock is usually a frequency of either 10 or 100 MHz. Legacy clocks may still provide 5 MHz, but 100 MHz is usually the norm today. By means of mixing and downconversion, this universal operation frequency is obtained. It may sound surprising, but most of these clocks do not provide a front panel, which actually shows time. The main emphasis is to have a very stable frequency source, which in the ideal case can be traced back to an atomic transition. There is no single and absolute setting for time. Time itself is a convention. So we can use the frequency of the clock generator and assign a time to a particular zero-crossing and count elapsed time up from that. The duration of a second is also a convention and this definition was there well before the cesium clock. This is the main reason, why it takes 9 192 631 770 oscillations for the ^{133}Cs isotope to last 1 s. Units of time such as 1 day, 1 h, 1 min and 1 s were defined to provide convenient subdivisions for our every day life. The clocks had to adjust to our needs of time. Although the indicated time scale may be wildly different from one clock to another, due to the definition of the length of a second, all clock would increment at the same rate.

A sine wave neither has a beginning nor an end. So, in order to make subdivisions of time, we need to detect a particular state and then start counting up cycles. Within one cycle we can use the phase and associate it with a particular time increment. For practical reasons it is very common to set the origin of time onto the zero-crossing of the sine wave, where the alternating voltage changes from negative values to positive values. This point has the steepest slope of voltage change. Therefore, this point is defined with the least uncertainty. The number of these time pips generated in this way can be counted in order to establish a time interval. Figure 2.27 illustrates this process graphically. The figure also shows another property of time and frequency very well. While frequency is a continuously smoothly changing waveform, time is given by distinct markers, which in an ideal case are infinitely sharp. The measurements of particular moments or events are defined by an epoch, which is a well-defined numbered instant on the respective time scale in use. Each clock realizes its own time scale and each time scale is only valid for how long the clock is continuously operating. When a clock is halted and then set into operation again, it also changes the respective time scale. The time pips generated from the clock oscillator are equidistant. They can be manipulated by subsequent numerical or digital electronic methods to code time, time intervals or timing codes. However, the critical component is always the clock oscillator. It defines how stable the clock runs if the time pips are truly equidistant. This is why the discussion of clock quality always is reduced to the underlying oscillator. In space geodesy we use time and frequency to:

- execute an event depending function and to record the epoch of a signal arrival;
- determine time intervals (e.g. time-of-flight measurements);
- count frequencies and generate frequencies for signal downconversion;

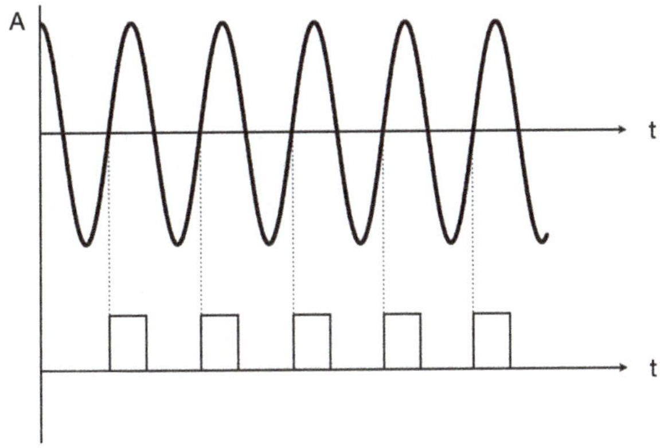

Figure 2.27. Illustration of the generation of time markers. In a technical realization he positive going zero crossing triggers a comparator, which then puts out a pulse. Counting these pulses then provides a measure for the elapsed coarse time.

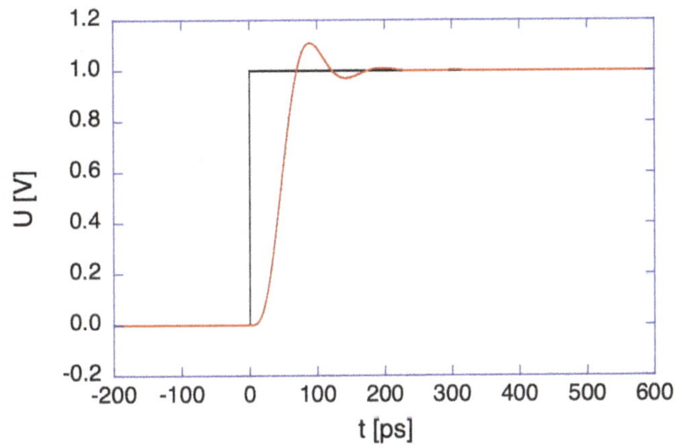

Figure 2.28. Illustration of the lowpass transfer properties of a voltage step function. The black curve switches from the low state to the high state within one sample. The lowpass filtered response function, shown in red, is delayed in time and rises much slower and takes time to stabilize on the new state.

- synchronize subsystems to a common source;
- generate and compare codes;
- facilitate time critical functions in measurement systems (e.g. steer a tracking telescope).

When we look at figure 2.27, we can see that the rising edge of the time markers needs to be as sharp in time as possible. While practically available laser pulses can be as sharp as 10 fs, electronic pulses over a cable are severely bandpass limited. Figure 2.28 illustrates this property. The input signal in this calculated example is a sharp step function that changes the state from one sample value to the next and is represented in black. The red curve presents this signal when it is lowpass filtered, simulating a passage through a bandwidth limited cable, albeit without damping. We can see that the response to the step function is delayed in time. Then the rising edge is no longer instantaneous, but a ramp with a given slope, which exhibits a near-linear behavior only in the middle part of the rising voltage. Finally, we see that it takes even more time before the voltage has settled on the new value and becomes stationary. In practical applications we can set a trigger somewhere on the near-linear part of the rising edge and still get stable timing conditions. However, this approach has to make sure that the bandwidth of the involved circuitry remains constant. Otherwise, we would obtain a variable delay problem.

The application of voltage comparators in the electronic circuitry of timing devices is very important. Therefore, we briefly introduce an example, outlining the principle of operation. Figure 2.29 presents a block diagram, where a timing pulse is entering from the left. Let us assume the signal was generated by a photomultiplier, hence the voltage is going negative. However, we note that this is an arbitrary choice. The photodetector in this envisaged application detects short pulses of light and these events are timed on a clock. When no signal is present the voltage output from the detector is 0 volts, corresponding to an idle state. When a signal arrives, the voltage line dips down by say −200 mV for about 100 ns. The leading edge marks the instant of the signal arrival. Although we would desire an instant response (with a slope of infinity) from our detector in order to time the moment of signal arrival with high temporal resolution, a real-world circuit does not provide this. Due to a limit in available bandwidth, we usually deal with signals that look like the sketched example in figure 2.29. Therefore, we need to set the trigger point V_{ref} to the steepest part of the falling edge of the pulse, assuming that the circuit always has the same temporal behavior and delay for all pulses as long as the incoming light pulse stays within a certain range of intensity, so that the detector does not saturate. The appropriate setting of the trigger threshold is indicated in the figure. When the signal voltage and the reference voltage are the same, the output of the operational

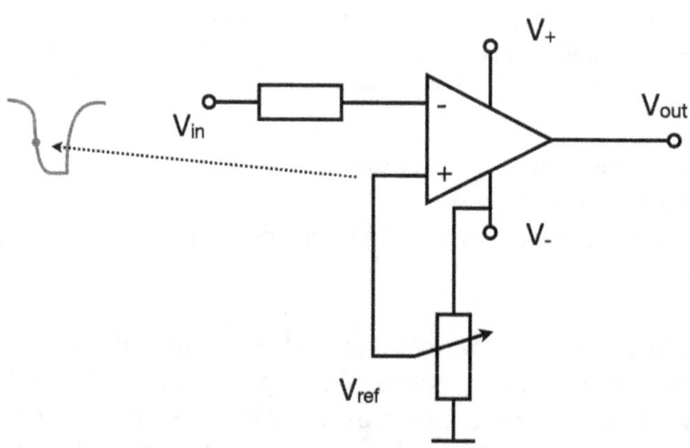

Figure 2.29. An electronic circuit diagram of an operational amplifier, acting as a comparator. When the incoming pulse signal voltage drops to a predefined value, the output voltage V_{out} flips polarity in an instant.

amplifier flips polarity from low to high and this voltage can now be used to latch a counter of timing pips from the clock oscillator and convert this number to the epoch of signal arrival. We cannot use the photomultiplier signal directly as the clock trigger because the signal line is noisy and operates on very small voltage changes. It also lacks the control to trigger the time latch at a definitive threshold of the leading negative edge. Digital circuits require higher signal levels and low noise to function properly. This concept of triggering and pulse shaping is very important in the measurement procedures in space geodesy, therefore we will look at some examples.

A *frequency counter* allows the accurate determination of the oscillation frequency of an unknown alternating voltage source; the functional concept is shown schematically in figure 2.30. A clock oscillator generates a gate signal of known length in time. The unknown alternating voltage passes through a signal conditioning stage, which converts the incoming frequency into a sequence of pips as seen in figure 2.27. When the gate voltage and the output of the signal conditioner are combined in a digital AND circuit and then counted, the frequency of the unknown signal source can be estimated. A frequency counter realized in this manner, however, has the inherent problem that it can only count the number of full oscillation cycles that fall into the gate window. As a result we have the uncertainty of $\delta = n/(n+1)$. By choosing a larger gate window, the number of cycles n can be made large and the measurement error reduces as $1/n$.

An inversion of the frequency counter concept is the *time interval counter*. Here we generate the signal pips from the internal known clock frequency and the gate open and close signal is generated from the external source. Usually we have one start input in the time interval counter and the stop input requires another signal connection. Figure 2.31 illustrates this concept. Here we use the frequency aspect of a clock rather than the time aspect in the previous example. The start and stop signal can have an arbitrary waveform. Therefore we require the appropriate trigger technique as described above. In terms of the measurement error the frequency counter and the time interval counter are similar. Increasing the clock frequency reduces the error, but has a practical limit, so that frequencies higher than 200 MHz are usually not available

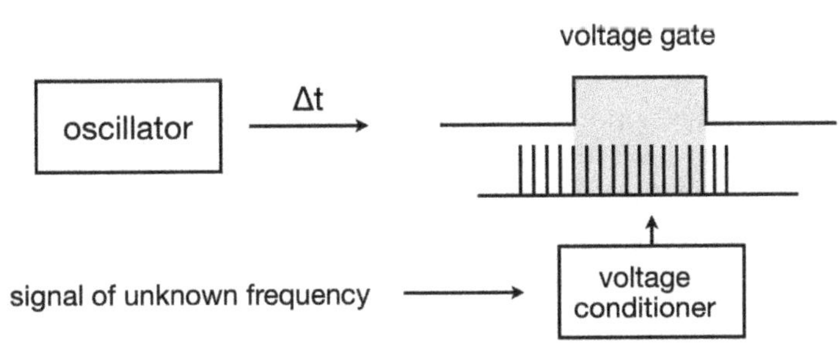

Figure 2.30. The functional concept of a frequency counter. A well-defined clock oscillator generates a gate voltage of specified length. The alternating voltage of an oscillator under test is converted to a sequence of pips in a conditioning stage. By a logical AND combination with a gate signal of known duration, the frequency of the unknown oscillator is established.

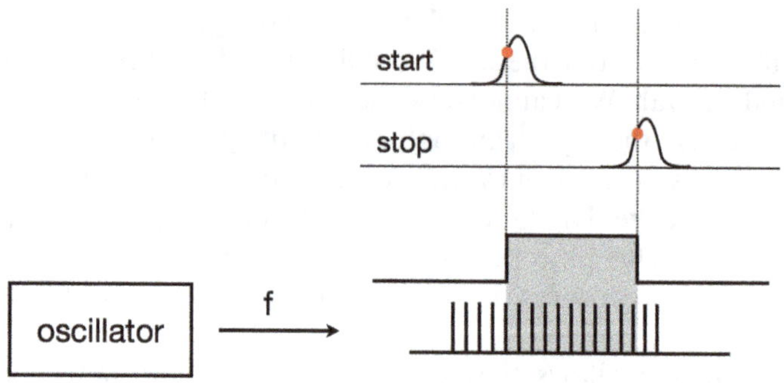

Figure 2.31. The functional concept of a time interval counter is the inverse of a frequency counter. The clock oscillator generates a sequence of known signal pips. These are counted in the interval between the start and the stop events, which set the gate.

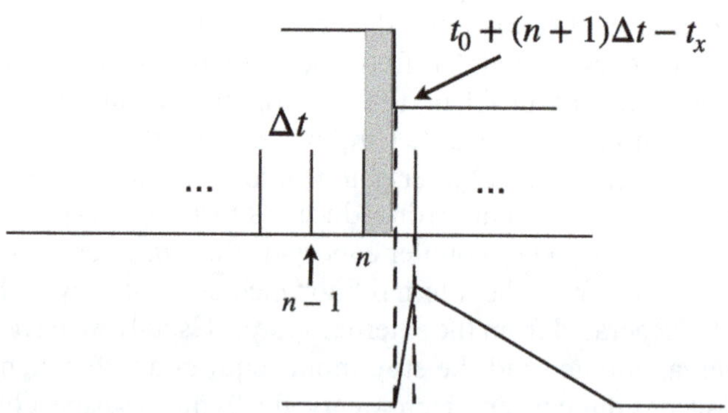

Figure 2.32. Principle of operation of a time-to-digital converter. The relative slow pip rate is counted directly. The remaining gray-shaded time segment is estimated from a linearized charge and discharge process. Since the discharge process takes much longer, this stretches the readout time and allows sufficient resolution to obtain a picosecond resolution.

because the separation from one pip to the next is already as short as 5 ns. The pulse width of each of these pips cannot be too short. Since the slope of the leading edge cannot be infinite, the signal voltage needs some time to respond. If there is not enough time available, the trigger point is never reached and the pip counting fails. Time interval counting is required for satellite laser ranging for example. If the technology would be limited to a 200 MHz time interval counter, one would not be able to measure distances with a resolution of shorter than 1.5 m in the round-trip or 75 cm in range. Required is a single shot resolution of 1 cm or preferably shorter. So this requires a different approach. One way to overcome this limit assumes that the highly time-resolved start or stop event detection is not required at a very high rate. This is a typical situation for the range-finding application. Such a time-to-digital converter operates in the following way (figure 2.32). A pip sequence that can be directly counted is generated from a clock oscillator. This is indicated by the vertical short lines in the figure. When a stop event occurs, it stops the counting process, which leaves the gray-shaded area as missing from the count range. The stop event, however, also starts a charging process on a small capacitor and the charging process is stopped by the next clock pip. This charging process utilizes a constant current, so that the charging process is linear with respect to time. We recall that this entire process is a matter of say 10 or 20 of nanoseconds. Furthermore, the acquired capacitor charge is proportional to the unwanted part of the last pip interval. When the charging process stops, the capacitor start to discharge. This process is also made linear in time but is made much longer than the charging process. This corresponds to a stretching of the discharge time relative to the charging time by something like a factor of one thousand and this can be measured by the clock rate with sufficient accuracy to provide a resolution of a picosecond for the gray-shaded area. The complete event epoch is then calculated by the initial time value t_0 plus $(n + 1)\Delta t$ the number of elapsed time pips—t_x, the short time interval derived from the capacitor circuit. It is important to note that in practice we estimate the small bit of one pip interval, which does not belong to the detected event and then subtract this from a full interval Δt.

Another way of extending the measurement range of an frequency counter is achieved by the mixing of two frequencies, where one is the unknown frequency source and the other is a known reference frequency. This technique is very universal and is available well into the GHz regime. Figure 2.33 illustrates this technique. The unknown frequency is amplified to sufficient signal levels, then it is superimposed with the known frequency from the reference oscillator. The result is a beat note as indicated in the block diagram. The envelope of the beat note is indicated in the diagram as well. One can also see that the higher frequency components are remaining in the interferogram. Therefore we require a lowpass filter in order to extract the down-converted frequency from the mix. The lowpass signal can then be counted and with the help of the known reference frequency, f_x can be calculated. Under normal conditions it is straightforward to obtain a valid result. Nevertheless we need to be aware of two important properties of this technique.

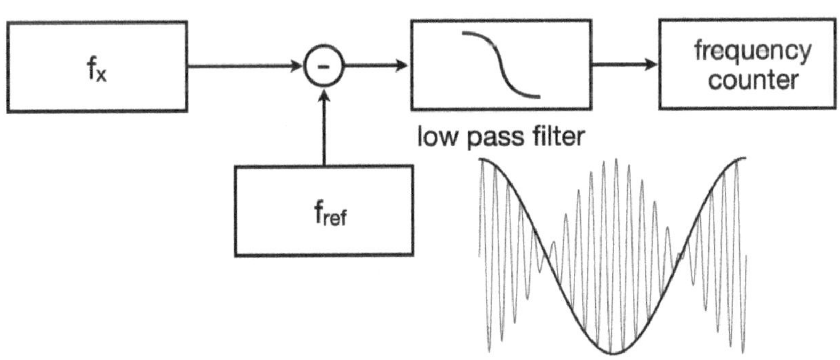

Figure 2.33. Mixing the high unknown frequency with a known reference into the range of the frequency counter extends the regime of operation. A lowpass filter is required to extract the down-converted frequency and avoids confusion for the counter stage.

1. The frequency mixing provides the absolute value of the difference between f_x and f_{ref}. This means that if the reference frequency is say 10 GHz, an unknown frequency of 10.2 GHz and one at 9.8 GHz produce the same beat note.
2. The stability and spectral purity of the reference oscillator can be crucial for the intended application, which is much wider than only frequency counting. The result is the combination of the properties of both signal sources.

The technique of downconversion with known reference frequencies is widely used in the techniques of space geodesy.

2.4 Historically important time scales

From a very general viewpoint we can define an arbitrary time scale by its origin, its unit of increment or scale and its direction of counting. In historical times before the realization of clocks, time was measured against astronomical observations, like day and night and the recurring seasons throughout the course of a year. Observations like the path of the Sun and Moon across the heavens led to subdivisions that were suited to the daily life. Early clocks, like a sundial supported this process. Early mechanical clocks followed this scheme as well. Old church tower clocks only had one hand, which indicated the hour. This in turn was a useful subdivision of a day at times where most of the daily activities were rather localized. The image of the church 'Rainkirche' in the city of Bruneck in northern Italy is one of many examples of this (figure 2.34). These clocks acted as interpolators between astronomically defined events. The clockworks were not very stable and therefore needed regular resynchronizations. It did not really matter how correct those clocks actually were. The processes of life were slow compared to our world today. This started to change noticeably with the advent of industrialization. With the invention of trains and industrial processing, life began to be less localized so that a finer subdivision of time was required and the minute hand was introduced to the clocks. This made better

Figure 2.34. Until today some historical churches still have clocks that only show the hour hand (left). Astronomical clocks also showed celestial configurations to mark the seasons (right).

clocks necessary. They still acted as interpolators relative to the rotation rate of the Earth, which at that time was considered to be an excellent and accurate natural clock. Around the transition from the 19th to the 20th century, time became the symbol of a modern society. Prominently visible attributes were the decorative pocketwatch, worn on a prominent silver chain by men and the grandfather clock, which decorated a lot of households. A typical tall pendulum clock had oscillation periods of the order of 2 s. Let us assume that the swing of the pendulum was stable to within 0.5%. This would mean that such a grandfather clock would lose or gain 7 min per day, despite the fact that this level of stability already marks quite an achievement.

In the days before telephone and public radio, this situation required some simple means for the public to adjust their clocks on a daily basis. The best way was the observation of the transit of the Sun across the local meridian. Figure 2.35 shows such an example. The photos were taken in the church Santa Maria degli Angeli e dei Martiri in Rome. While the solar transit in Rome was used to mark cycles in the Christian year, the city of Prague used the same concept for public clock synchronization and utilized what was known as the *lunchtime shot*. When the Sun crossed the meridian in the Klementinum in Prague, a flagman gave a signal from the rooftop of the tower, where the meridian was located. This prompted a shot from a cannon from one of the nearby hills. This shot was heard in all of Prague and every household could set the clock. On a rainy day, when the solar transit was not observable, the time was taken from an ensemble of high-quality mechanical clocks. These were much more stable than any commercial grandfather clock. While using the solar transit across the local meridian provided a good way of adjusting a clock, this method has an inherent problem. The local meridian provides a reference to midday from the rotation rate of the Earth only in the respective location. At the equator the Sun moves almost 1667 km per hour. At midlatitudes this reduces to 1239 km. This means that the time difference for noon between the cities of Prague and Nuremberg, which are 276 km apart already amounts to 13.4 min. While this

Figure 2.35. A hole in the roof acting as a pinhole allows the Sun to cast a mark on the floor of the church, where a meridian line is integrated.

did not have any negative consequences for all local purposes, it made life difficult for large-scale train operations. Therefore, it is no surprise that time zones were introduced at the end of the 19th century. These local clock synchronizing practice came out of fashion when better clocks and other means of synchronization became widespread. Quartz clocks, for example, improved the stability by several orders of magnitude and made frequent clock synchronizations unnecessary. The radio transmission of time markers was introduced when radios became generally available. Using the radio, telegraph or the telephone for setting the clock was also much more accurate. At the same time the need for a uniform time scale covering a large area became more and more of a necessity.

Precise timing was also important for sea travel. This also required some technique for clock synchronization. Audio signals, like the shot of a cannon, were not accurate enough for this purpose. Since the distance between the local master clock and the various ships was too long, the travel time of the sound was already significantly limiting the achievable accuracy. Many harbors adopted a timeball instead. Figure 2.36 shows the timeball station of Lyttleton Harbor in New Zealand as an example. Shortly before noon, the timeball was pulled up the post. When the appropriate time signal was received by the telegraph, it released a catch and the timeball dropped from the top of the post. This was noticed by the navigator of the ship and he triggered the clock synchronization process. Again, this process was discontinued when better clocks became available.

The beginning of the new day is also a matter of convention and different choices were made throughout history. Sunrise or midday appear to be a natural choice, but sunset and midnight were also favored options. The disadvantage of sunrise and sunset in Western Europe is the fact that the times of sunrise and sunset change throughout the year as the Sun cycles through the seasons. Midday has the advantage that it could be well measured, but it was not convenient for daily life to start the count of a new day then. Nevertheless, astronomers and navigators preferred the choice of the change of day at noon because it provided more accurate

Figure 2.36. The timeball station in the harbor of Lyttleton, New Zealand. When the noon signal was received by telegraph, it dropped the ball from the top of the post.

results [13]. The Julian day numbers, starting at 4713 BC, began the day at noon and so it was for the Greenwich Mean Time, until this changed to midnight for the beginning of the new day. The shift from noon to midnight was progressively introduced from about 1925 [13]. By 1930 this transition was generally accepted and consistently used. Time is a good example to illustrate that the definitions of measurement systems and units are essentially triggered by changing societal needs. These needs can be economical, religious or scientific. For time we can summarize that until 1956 the length of a second was defined as

$$1\ [\mathrm{s}] = \frac{1}{86\,400} \tag{2.31}$$

of the mean solar day, where the solar day is the time between two successive transits of the Sun over the local meridian. Due to small irregularities in the rotational motion of the Earth, the length of the solar day varies slightly and has to be averaged out. We note again that the approximate length of a second was set by needs and had been agreed upon before that. This definition was derived from astronomical observations in order to make it precisely reproducible. Between 1956 and 1967 this definition turned out to be not good enough. The rotational motion of the Earth was influenced too much by mass transport phenomena of the liquids of the Earth, namely water, atmosphere and cryosphere. So it was superseded by another astronomical observation. This time the orbital velocity of the Earth around the Sun was used, where mass transport does not matter. The second then became

$$1\ [\mathrm{s}] = \frac{1}{31\ 556925.9747} \tag{2.32}$$

of the tropical year in the year 1900. This is the period of the Sun passing through the same position in successive times, for example from one spring equinox to the next. A tropical year has 365.2422 solar days or 366.2422 sidereal days. A solar day is the time between two successive transits through the local meridian and contains a fraction of the Earth's orbit around the Sun, through which the Earth has passed within that day. The sidereal day is the time between two successive transits of the same star over the local meridian. This observation does not contain the motion of the Earth along its orbit. So the sidereal day is 3 min and 56.6 s shorter. For the length of the year of n = 365.2422 days have obtain the length of the sidereal day as

$$t = 86\,400\frac{n}{n+1}\ [\mathrm{s}] = 86\,164\ [\mathrm{s}]. \tag{2.33}$$

In 1967 this definition was substituted by a fairly radical step. In the mean time the development of clocks based on an atomic transition had matured to the point that they became much more stable and reliable than the astronomical observations. The new definition selected the cesium clock (discussed in section 2.2.2) as the new standard because of the superior long-term stability. Today the duration of one second is defined as

$$1[\mathrm{s}] = 9\ 192\ 631\ 770 \text{ oscillations of the } {}^{133}\mathrm{Cs} \text{ isotope}. \tag{2.34}$$

In the year 1999 this definition was amended with the statement that the definition requires the Cs atom to be at rest near 0 K. This criterion is generally met in Cs-fountain clocks. We remember now that optical clocks have improved significantly over microwave clocks. There has been an ongoing discussion for years about when the definition of the second will be based on an optical transition. At the time of writing the microwave clocks operate continuously over a long time, but optical clocks only run intermittently. When the redefinition of the second is required, it will happen after a sufficient time of comparing the new standard with the old one. Right now it appears that there could be a secondary definition of the second based on an optical clock concept within a decade. This includes the consideration of the difficulty of comparing optical clocks reliably as well. However, this decision will turn out, the redefinition of the second from one atomic standard to another is far less radical than the transition from an astronomical to a purely physical definition.

The observation of the position of the Sun in the sky provided a practical measure of time for many centuries. If we compare midday to a good local clock over the course of a year, we will notice that the days are not uniform in length. The path of the Sun is inclined with respect to the equator and the Earth is not in a circular orbit around the Sun. The consequence is that the length of a solar day varies. We observe the apparent solar time. If we compare the apparent solar time against the mean solar time, we obtain the equation of time, which is shown in figure 2.37. The differences can accumulate to as much as half an hour between the extreme positions. The globally used time scale is the *Universal Time* (UT), which is given by the mean solar time at the location of the Greenwich meridian. And this is also the origin of the longitudes representing the value 0. This time scale does not contain polar motion, seasonal and irregular variations in the rotation rate of the Earth. Therefore the resolution of the UT time scale is limited to $\delta T/T = 10^{-8}$. The application of clocks based on an atomic transition allows for a much better

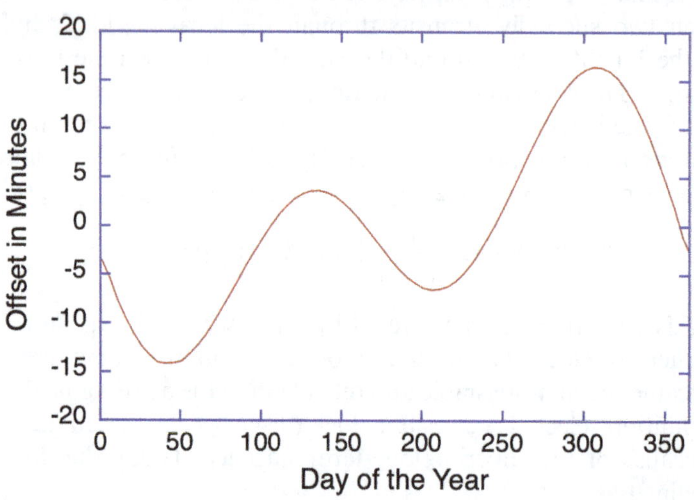

Figure 2.37. The equation of time is given by the difference of the mean and the apparent solar time.

resolved time scale. So we can distinguish several levels of correction of the UT time scale.

- UT0 = UT.
- UT1 = UT0 + corrections for polar motion.
- UT2 = UT1 + corrections for seasonal variations.

Seasonal variations are caused by mass relocation of fluids (ice sheets, water and atmosphere) of the Earth, which follow a repeating annual pattern. They cause a variation of the moment of inertia of the Earth, which is similar from one year to the next. The equation of time, the precession and nutation motion of the rotational axis of the Earth and the tidal friction are already contained in UT0. The corrections between UT1 and UT0 amount to values smaller than 70 ms.

Mass redistribution of the fluids on the Earth causes some variability with respect to the rotational velocity of the Earth. We have seen this in the paragraph above. However, the orbital velocity of the Earth can be derived with higher accuracy because the total mass itself does not change and all other gravitating objects are far enough away, so that this does not significantly interfere with the Earth's orbit. Therefore, we can obtain a precise time scale from the calculation of the ephemeris[4] of the Earth's orbit. The ephemeris time was introduced in 1960. This led to the definition of the ephemeris second as 1 part in 31 556 925.9747 of a tropical year for the epoch 1 January 1900 at midday. The tropical year corresponds to a complete revolution of the Earth around the Sun. Taking it very exactly, we find that the length of the tropical year has a slight dependence on the eccentricity of the Earth as well as on the precession and nutation of the Earth. The length of the tropical year is determined to be 365.2422 days long (after Newcomb) astronomically. The Earth's orbit is not very convenient for a time scale because it takes a long time and a lot of observations to establish this time scale. Another option is given by the lunar orbit, which is about an order of magnitude faster to establish. the ephemeris time scale can be established with a residual uncertainty of $\delta T/T \approx 10^{-9}$. This is one order of magnitude more precise than the UT time scale. Today the ephemeris time scale only has historical value, since the atomic time scale is much more stable. Furthermore, due to the higher frequency of oscillation it is also a lot easier to subdivide the atomic time scale. That means that the atomic time scale also has a much better resolution.

The date and the time run in cycles. For example, the count of the minutes in an hour roll over from 59 to zero at the full hour. The same applies for the seconds in a minute and the hours in the day. Finally even the count of the days are not continuous. They roll over at the change of a year. For many applications and in particular for the generation of a long time series this is very inconvenient. In these cases we require a continuous segmentation of units of time. The *Julian date* (JD) is a commonly used method to achieve this goal. The beginning of the Julian time scale is at noon of January 1, of the year 4713 BC. This count increases by an integer of 1 from one day to the next. Within a day, the time is given in decimal fractions, which can be calculated

[4] The ephemeris is the calculated orbit position of a celestial object in the sky.

by the quotient of the elapsed number of seconds in the day and the total number of seconds per day, which is 86 400 by definition. While the Julian date provides the convenience of a continuous time scale, it has two disadvantages. First of all, it starts at noon, which is owing to the fact that noon could be established more easily. Secondly, there are a lot of digits to work with. This led to the definition of the *modified Julian date* (mJD), which can be derived from the Julian date by subtracting the constant value 2 400 000.5. This shifts the change of the day to midnight and effectively drops the first two digits in order to make numbers more handsome. In this convention the first of May 2025 becomes in the mJD conversion day number: 60 796. We are using the mJD convention for ring laser analysis later in this book.

In this chapter, we have realized that there are as many time scales as there are clocks. In order to compare clocks, we need to come up with a time scale, which does not depend on a physical clock and which is globally valid. Furthermore, we need to make a distinction between a time scale that is purely defined by atomic clocks and a time scale, which is adjusted to the rotation rate of the Earth. We refer to this latter time scale as *Universal Time Coordinated* (UTC) and the former time scale as *Temps Atomique International* (TAI). These time scales are defined in the following way:

- Definition of the origin: 1 January 1958, 0:00 o' clock UT2.
- Scale: the SI second.
- Direction of counting: ascending, where the UTC and the TAI. time scale only differ by an integer number of seconds (leap seconds).

Highly resolving technical clocks today usually provide a reference frequency of 100 MHz and electrical square-wave tick-marks, indicating the beginning of the second of the particular clock. These tick-marks are usually referred to as the *pulse per second* (PPS) signal. In order to compare clocks, one has to measure the offset between the PPS signal of the two clocks under consideration. Figure 2.38 illustrates the process. The clock comparison can only be done at the velocity of light, and

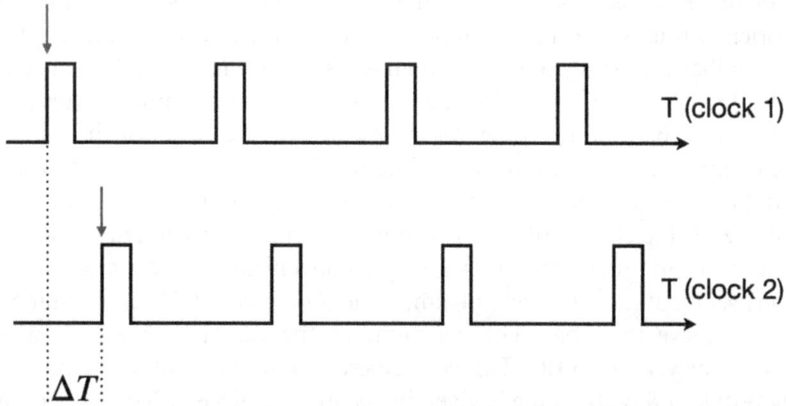

Figure 2.38. Illustration of the clock comparison process. The beginning of the second is the rising edge of the PPS tick-mark. The offset ΔT is typically in the millisecond range.

while this is fast, it is still only 30 cm per nanosecond and we need to define where on each clock the origin of the respective time scale is. For all practical purposes it is enough to define one particular PPS output port as the point of reference on each clock. Furthermore, we need to either measure the delay of each cable used in the comparison process, or to make sure that they are identical. We then get for the difference in time of the two rising edges of the PPS

$$\Delta T \big|_1^2 = T_2 - T_1. \tag{2.35}$$

If the offset between the two PPS signals remains constant over time, both clocks are running on the same value for the scale. This is very rarely the case. Usually we will find that the offset will not be constant. That means that there is an error in scale of one of the clocks, or both. We can estimate this, but since the difference is normally very small, we cannot say which of the clocks developed the error in scale. To estimate this error by measurement we take

$$\frac{\Delta f}{f} = \frac{\Delta T \big|_{1T2}^2 - \Delta T \big|_{1T1}^2}{T2 - T1}. \tag{2.36}$$

It takes at least a third clock to establish, which of the clocks in question has developed the error in scale. Under normal conditions all of the clocks have small errors in scale and the clock with a gross error will be identified quickly. Each timing laboratory usually has a fairly large clock ensemble, to ensure that it can keep time. All these clocks are continuously compared to each other.

The Geodetic Observatory Wettzell operates three cesium clocks, three hydrogen masers and a cold atom μclock and is one of four timing laboratories in Germany. The main time laboratory is at the national metrology institute: 'Physikalisch Technische Bundesanstalt' (PTB) in Braunschweig. The other two are at the 'Deutsche Telekom' and the 'Deutsches Zentrum für Luft- und Raumfahrt' (DLR) in Oberpfaffenhofen. One of the clocks in the ensemble is defined as the master clock. The other clocks are compared against the master in order to verify the integrity of the master clock. The clock frequency of the master can either be 5 MHz, 10 MHz or 100 MHz, depending on the respective manufacturer. In our case the 5 MHz frequency is distributed to the geodetic measurement systems, say, for example, the SLR system. This frequency is a narrow-band signal and therefore it is not significantly degraded by a long cable run. At the SLR site the clock frequency is fed into a clock module, which generates the PPS pulses of the local SLR time scale. These PPS pulses are fed back to the timing laboratory. This time the square-wave pulses of the PPS signal are broadband timing signals, and the long cable, acting as a lowpass filter, rounds off the leading edges of the PPS pulses. This causes a small uncertainty for the exact estimation of the beginning of the second of the PPS. Moreover, there is also a significant delay introduced by the length of the cable, which has to be established and to be applied as a clock offset correction numerically by adding this delay to the recorded epoch of the geodetic measurement. Figure 2.39 shows a simplified block diagram of the time distribution setup for just one of the measurement systems. The involved total cable runs for time at the observatory have a length of the order of 100—200 m per measurement system. They

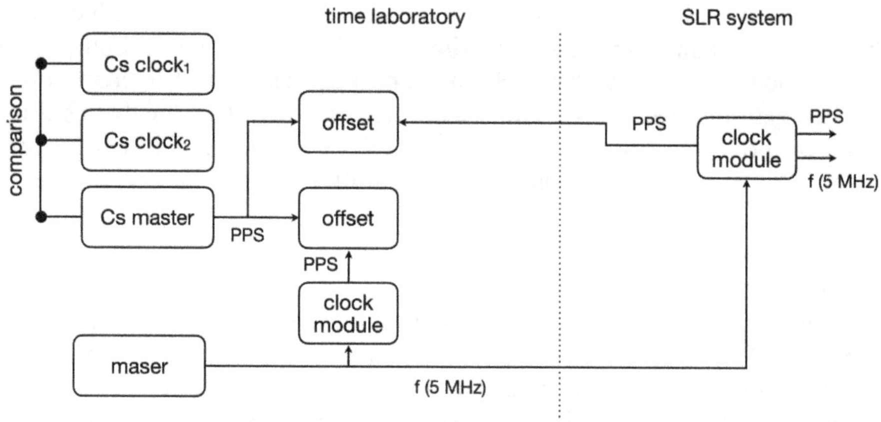

Figure 2.39. A block diagram for the time distribution process on the Geodetic Observatory Wettzell. The maser frequency is used, while the actual time is provided by an ensemble of Cs clocks. The PPS generated at the local SLR system is fed back to the time laboratory in order to establish the clock offset between the two locations.

start at the time laboratory and go to the SLR and the VLBI systems. In practice it turns out that the delay error can be maintained to stay within a 10 ns window over a full year. A synchronization error of this size is fully acceptable for the VLBI and SLR measurements. This is discussed together with the respective techniques in later chapters. Here we still need to attach our time laboratory to the UTC time scale.

Since the observatory in Wettzell operates a time laboratory, it realizes the local time scale UTC_{IFAG} and contributes to the UTC time scale by clock comparison with the *Bureau International des Poids et Mesures* (BIPM), located in Paris (France). The time department of the BIPM publishes the Circular T on a monthly basis [14]. The Circular T lists clock and clock comparison data of approximately 80 participating time laboratories, most of which are operated by national institutes. It provides traceability between UTC and the local $UTC_{(k)}$ realizations by reporting [UTC − $UTC_{(k)}$] on a five-day basis, where (k) is the particular time laboratory; in our case, (IFAG). Figure 2.40 provides an example. UTC_{IFAG} is based on three different commercial HP5071A cesium clocks. Utilizing daily GNSS ‘common view’ or ‘all in view’ clock comparisons with the PTB and the BIPM provides the respective offsets between Wettzell and these timing laboratories. Taking a closer look at the Circular T, we can see that the determined offset between the BIPM and many time laboratories is fairly small and in the range of either single- or double-digit values of ns. This usually indicates that these clocks are steered toward UTC. Since it is not a good idea to upset a clock by shifting the rising edge of the PPS tick, these laboratories usually feed the cesium clock signal to a phase-shifting device in order to move the beginning of the second closer to UTC. It is a delicate process because each clock obtains a weight for the estimation of UTC, which is based on the performance over the previous month. If the offset shifts a lot, the weight of this clock for the coordinated time scale is low. For the UTC_{IFAG} we find a rather large value for the offset in the list. The clocks in Wettzell

CIRCULAR T 446 ISSN 1143-1393
2025 MARCH 11, 07h UTC

BUREAU INTERNATIONAL DES POIDS ET MESURES
THE INTERGOVERNMENTAL ORGANIZATION ESTABLISHED BY THE METRE CONVENTION
PAVILLON DE BRETEUIL F-92312 SEVRES CEDEX TEL. +33 1 45 07 70 70 tai@bipm.org

The contents of the sections of BIPM Circular T are fully described in the document "Explanatory supplement to BIPM Circular T" available at https://webtai.bipm.org/ftp/pub/tai/other-products/notes/explanatory_supplement_v0.8.pdf

1 - Difference between UTC and its local realizations UTC(k) and corresponding uncertainties.
From 2017 January 1, 0h UTC, TAI-UTC = 37 s.

Date 2025 0h UTC	JAN 29	FEB 3	FEB 8	FEB 13	FEB 18	FEB 23	FEB 28	Uncertainty/ns			Notes
MJD	60704	60709	60714	60719	60724	60729	60734	uA	uB	u	
Laboratory k				[UTC-UTC(k)]/ns							
AGGO (La Plata)	458.7	452.1	452.9	445.6	459.8	446.8	427.9	0.7	2.9	3.0	
AOS (Borowiec)	-2.8	-2.6	-2.0	-1.5	-0.4	-0.8	-0.8	0.2	2.6	2.6	
APL (Laurel)	0.0	0.5	0.7	1.1	1.8	1.4	1.3	0.2	NC	-	
AUS (Sydney)	73.3	97.0	107.0	119.2	99.2	68.6	17.5	0.2	2.9	2.9	
BEV (Wien)	-6.6	-7.4	-3.0	1.1	5.3	7.4	8.7	0.2	2.9	2.9	
BFKH (Budapest)	1278.9	1328.9	1387.7	1435.3	1491.5	-	-	0.7	7.1	7.1	
BIM (Sofiya)	3101.9	3150.1	3193.8	3237.7	3281.9	3333.2	3378.5	0.2	2.6	2.6	
BIRM (Beijing)	1.5	0.6	-0.5	1.0	3.0	1.1	2.5	0.2	3.5	3.5	
BSJ (Kingston)	33.8	13.1	14.6	31.7	20.1	20.7	26.6	7.0	7.1	10.0	
BY (Minsk)	-0.8	-0.6	-0.6	0.4	1.7	1.5	1.3	1.5	3.1	3.4	
CAO (Cagliari)	-19117.7	-19232.4	-19349.8	-19471.0	-19591.9	-19714.2	-19835.6	0.7	NC	-	
CH (Bern-Wabern)	-1.9	-1.2	-0.4	-0.1	-0.6	-0.4	0.1	0.2	1.0	1.0	
CNES (Toulouse)	3.7	3.0	1.2	-1.4	0.4	1.6	2.3	0.2	3.1	3.1	
CNM (Queretaro)	2.9	0.2	-0.6	-0.3	-0.3	-0.1	-1.0	3.0	4.4	5.3	
CNMP (Panama)	-6.3	-18.9	0.6	6.5	4.2	6.1	4.5	0.2	5.5	5.5	
DFM (Horsholm)	62.4	67.7	73.4	78.6	59.1	-0.7	-0.9	0.2	2.7	2.7	
DFNT (Tunis)	-7474.4	-7570.0	-7653.8	-7742.0	-7834.8	-7929.9	-8016.5	0.7	2.6	2.7	
DMDM (Belgrade)	14.0	18.2	28.8	32.1	46.3	46.2	52.0	0.2	7.4	7.4	
DTAG (Frankfurt/M)	16.5	11.1	9.8	0.5	0.4	-1.1	-1.6	0.2	3.4	3.4	
ESA (Noordwijk)	0.8	0.7	-0.3	-0.5	0.1	-0.2	-0.3	0.2	2.8	2.8	
HKO (Hong Kong)	1225.0	1256.6	49.4	53.0	54.6	49.0	57.3	0.4	3.6	3.6	(1)
IBM (La Paz)	-	384.5	378.6	368.2	356.7	367.7	356.5	4.0	7.6	8.6	
ICE (San Jose)	-25.6	-39.4	-41.0	-34.2	-35.7	-33.1	-28.0	0.2	7.1	7.1	
IDN (Serpong-Tangerang)	-672.7	-674.6	-677.1	-692.7	-712.2	-717.7	-719.3	0.2	NC	-	
IFAG (Wettzell)	-708.1	-706.9	-702.8	-690.6	-687.6	-682.3	-680.2	0.2	3.1	3.1	
IFZG (Zagreb)	66.6	62.6	67.9	69.3	73.8	80.7	88.4	0.2	2.6	2.6	
IGNA (Buenos Aires)	-	-	-	-	-	-	-				
IMBH (Sarajevo)	6.3	-2.0	0.7	1.9	3.0	3.5	6.1	0.2	2.8	2.8	
INCP (Lima)	-39.6	-45.0	-45.4	-36.5	-39.6	-29.5	-30.2	5.0	NC	-	
INM (Bogota D.C.)	-21.5	-28.5	-28.4	-14.6	-18.3	-3.5	-41.6	3.0	NC	-	

Figure 2.40. An extract from the Circular T published for February 2025 by the BIPM showing the values of [UTC − $UTC_{(k)}$] in the units of ns. The given estimates are each 5 days apart.

are not steered at all. They are operated as free-running clocks, which are tuned only when the offset becomes prohibitively large. Again, it is not necessary to perform the measurements of SLR and VLBI with epoch estimates, which are very close to UTC. An offset in the range of about 1 μs is not of concern. Figure 2.41 shows the performance of the free-running cesium master clock over a period of 5 years, namely between 8 July 2006 and 1 August 2011.

Now we still have to examine how UTC itself is generated. This is quite an involved process. We have seen already that earth rotation and the TAI time scale generated by atomic clocks differ due to irregularities in earth rotation. Between the beginning of TAI in 1958 up to the end of 1999 a difference of 31.3 s out of a total of 1 325 376 000 elapsed seconds had developed. This corresponds to a rotation angle offset of 0.13° [1]. We remember that the UTC time scale is tuned to earth rotation by inserting a leap second when the offset increases beyond 0.7 s. The block diagram in figure 2.42 shows how TAI and subsequently UTC are generated. Several NMIs operate primary cesium frequency standards. This means that the respective cesium clocks and more importantly the cesium fountain clocks of these institutions have a fully characterized and sufficiently small error budget and a small instability. Therefore, these primary clocks from institutions in France, Japan, Russia, USA

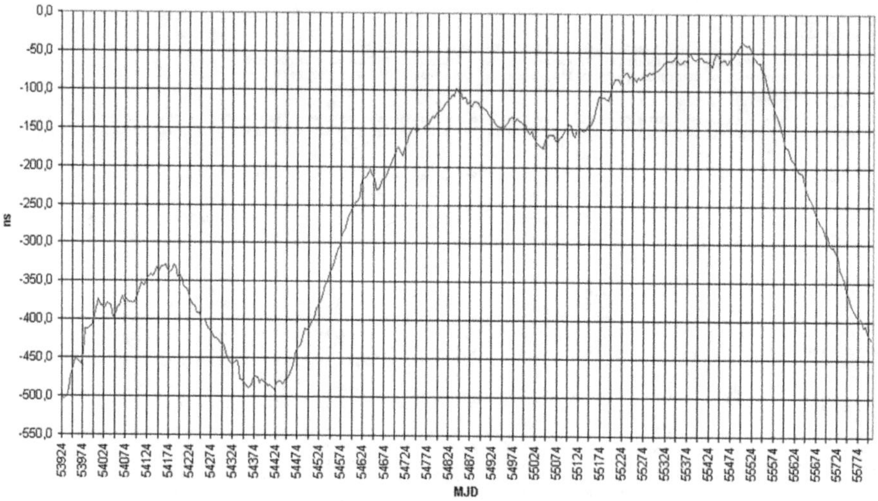

Figure 2.41. The illustration of the drift behavior of the cesium master clock in Wettzell over a period of 5 years. The offset stayed within a window of 500 ns.

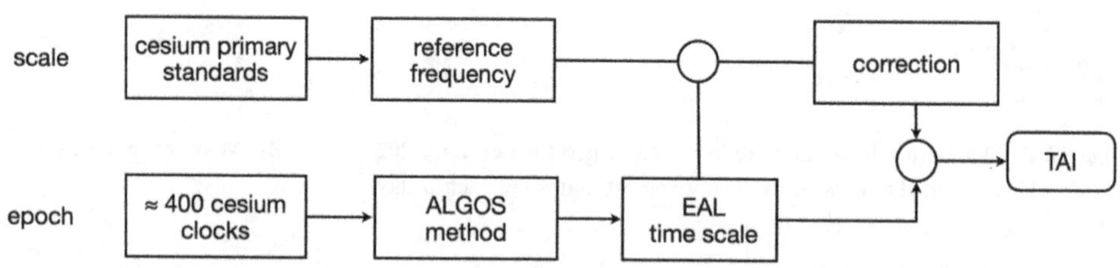

Figure 2.42. Block diagram of the composition of TAI. The scale is obtained from the primary standards, while a large ensemble of commercial clocks from many time laboratories are fixing the actual time, weighted according to their estimated instability from the clock comparison of the previous month.

and Germany are used to define the scale of the SI second. In order to obtain a time scale, a large number, approximately 400 commercial cesium clocks from many different time laboratories are evaluated. The behavior of these clocks is known from the frequent time comparisons. An involved evaluation algorithm (ALGOS) was developed for this purpose. It weighs each clock according to its determined instability over the last month. The goal is to provide a reliable and stable time scale from this derived mean value of all contributing clocks. This intermediate product is called *Échelle Atomique Libre* (EAL) and constitutes a free atomic time scale. It still lacks a correction for the proper scale. Once this adjustment has been carried out, we obtain the TAI time scale. It sounds a bit curious, but the TAI time scale is only fixed retrospectively, once the adjustment process has been performed. However, for all practical purposes this does not cause a problem. From the theory of general relativity we know that gravity influences time. So we need a correction for the physical height of the clock. This is discussed in section 2.5.

2.5 Relativistic corrections

The notion of time depends on the reference frame in which it is derived. Very soon we stumble over the expression 'proper time'. Very intuitively this would be the time that I read from a clock that I have on me. More generally, the proper time in general relativity is the time measured by a clock that follows a timelike worldline, which sounds rather abstract. In practice it means that we can compare clocks by putting them side by side and read them simultaneously. If the clocks I wish to compare are in motion relative to each other or they are not in the same place, then we need to apply corrections. Let us list some facts.

- The process of taking time depends on the frame of reference. Each reference frame has its own time.
- Taking the time does not depend on the type of the utilized clock. A hourglass and a cesium standard exhibit the same properties, but not the same resolution of course.
- A clock comparison tests the structure of the laws of physics. The experienced effects, like for example time dilation are the same for all clocks.
- The behavior of clocks is universal, but the geometry involved in the clock comparison may require corrections.

Clocks may move with respect to each other during the comparison process and they may be in locations of different gravitational potential. Therefore, we require a set of rules in order to compare or synchronize them. Highly accurate clock comparisons of remote clocks make implicit use of the **isotropy** of the speed of light, which is the property that in vacuum the speed of light is the same in all directions[5]. In order to compare remote clocks, we can use what is known as the **Einstein synchronization**:

> Two different clocks A and B, which are at rest in an inertial system can be synchronized by pulses of light. The light pulse transmitted from location A at time t_1 arrives at location B at the time t_2. It time-tags the clock at location B and gets reflected without delay by a mirror at the location of the clock. It returns to location A at the time t_3. Clock B runs synchronously with clock A when we have
>
> $$t_2 = t_1 + \mu(t_3 - t_1) \quad \text{and} \quad \mu = \frac{1}{2}. \tag{2.37}$$

This is a very generally formulated rule, but we can immediately recognize that satellite laser ranging is the technical realization of the Einstein synchronization process. Furthermore, it also provides the fundamental concept of optical time transfer. We also recognize that the process of remote clock comparison takes time itself, which needs to be removed from clock comparison. The second term $\frac{1}{2}(t_3 - t_1)$ in equation (2.37) does

[5] Although the isotropy of the speed of light is taken as granted in space geodesy, it is still subject to validation in fundamental physics. However, any violation of this rule would be too small to matter for our purpose here.

this for us. Similarly important is the fact that the highest possible velocity that can carry information is the velocity of light c. Actually one does not need to measure c. Today c is one of the physical constants and amounts to 299 792 458 m s^{-1}. It relates to other fundamental constants as

$$c = \frac{1}{\sqrt{\varepsilon_0 \mu_0}}, \tag{2.38}$$

where ε_0 is the vacuum permittivity and μ_0 the vacuum permeability. This property has a profound effect on time transfer and orbit determination. It enables SLR, VLBI and GNSS as well as time transfer. We can measure distances with a resolution of mm and transfer time at the level of a few ps. Although the measurement process is very simple, the technical realization nevertheless is very involved and requires a significant technical effort.

If the clocks of interest are located in two different reference frames, which move with respect to each other, we need to apply a correction according to the special theory of relativity to account for the finite value of c.

Two inertial systems I and I' in motion with respect to each other with two events E and E', which are given by the coordinates X and X' and the respective times t and t' are linked to each other by the **Lorentz transformation**:

$$X' = \frac{X - vt}{\sqrt{1 - \frac{v^2}{c^2}}} \quad \text{and} \quad t' = \frac{t - \frac{vX}{c^2}}{\sqrt{1 - \frac{v^2}{c^2}}} \tag{2.39}$$

In equation (2.39) we denote the relative velocity between the two moving clocks as v. One of the important consequences of the finite velocity of light is the fact that events that are simultaneous in the inertial system I, are not simultaneous in the inertial system I' [4, 5]. This is distinctly different from Newton's mechanics, where time and space were absolute. For the difference in time between two events separated by a distance we now get

$$\Delta t' = \frac{\Delta t - \Delta x \frac{v}{c^2}}{\sqrt{1 - \frac{v^2}{c^2}}}. \tag{2.40}$$

The expression in equation (2.40) simplifies if the two events are in the same spot:

$$\Delta t' = \frac{\Delta t}{\sqrt{1 - \frac{v^2}{c^2}}} \tag{2.41}$$

Table 2.2. Example for the observed time dilation depending on the relative motion of the observer.

Object	Time dilation [s]
Plane (1000 km h^{-1})	0.000 000 000 000 034 (0.034 ps)
Meteorite (34 km s^{-1})	0.000 000 000 180 (180 ps)
10% of the velocity of light	0.000 050 4 (50 μs)

This has a very general consequence, namely:

Because of the time dilation a moving clock runs slower than a clock at rest.

This aspect of special relativity as expressed in equation (2.41) is always a bit counter-intuitive. Therefore, we look at an example (table 2.2). A plane traveling at a velocity of 1000 km h^{-1} passes a spot where two lamps are switched on with a delay of $\Delta t = 0.1$ s. What delay would the pilot see? What delay would an observer passing the same spot on a meteorite of 34 km/s see, and finally, if you are passing the spot at 10% of the velocity of light, what would the delay be then? When we compare two clocks, one in a reference frame at rest and the other in an accelerated reference frame in the time interval from t_1 to t_2, the difference in time accumulating:

$$\Delta t' = \int_{t_1}^{t_2} \mathrm{d}t' = \int_{t_1}^{t_2} \frac{\mathrm{d}t}{\sqrt{1 - \frac{v(t)^2}{c^2}}} \tag{2.42}$$

The corrections so far have been a consequence of the special theory of relativity. In space geodesy we also have to consider an effect from general relativity, which is related to gravity and known as **gravitational redshift**. For a time interval Δt on a clock located on the geoid at $h = 0$ and an identical clock at the height h we see the time interval $\Delta t'$ with the relationship

$$\Delta t' = \left(1 + \frac{gh}{c^2}\right)\Delta t. \tag{2.43}$$

We can conclude:

A clock at a higher location is running faster than a clock at a lower location.

TAI is defined in a geocentric frame of reference. That means the clocks are referenced to the geoid or more generally to the sea level. However, the clock laboratories are usually not at sea level. Therefore, a correction for the gravitational redshift has to be applied. This correction is -1.09×10^{-16} per m [1]. This implies that the PTB clocks at $h = 78$ m above sea level run faster than the SI second by a factor of -8.6×10^{-15}.

With the advent of highly resolving optical clocks in the last few years, there have been quite a few highly sophisticated experiments to exploit this effect [11, 19] by comparing optical clocks between Paris and Braunschweig and utilizing two Strontium lattice clocks to infer the height difference of the clocks at the top and the bottom of the broadcasting tower 'Tokyo Skytree'. For the 452.65 m difference in height, a redshift of the optical clock frequency of 21.18 Hz was found in accordance with Einstein's prediction.

Another spectacular experiment for the test of the theories of relativity was performed by J C Hafele and R E Keating in 1972. They packed a set of four cesium clocks onto an aircraft and flew around the world, one time in the west direction and one time in the east direction. Before takeoff and after the landing the traveling clocks were compared to the clock ensemble at the US Naval Observatory in Washington [6, 7]. To make sure the clocks were performing well, they were constantly compared among each other during the journey. Figure 2.43 illustrates the experiment. The necessary corrections from the various contributions are:

$$\Delta t = \int_{\text{path}} \left(1 - \frac{g(\varphi)h}{c^2} + \frac{v^2}{2c^2} + \frac{\omega a_1 v_E \cos\varphi}{c^2}\right) \mathrm{d}\tau \tag{2.44}$$

There are three terms to consider in equation (2.44). The first term provides the gravitational redshift contribution. The second term accounts for the transport velocity and the third term corrects for the effect of earth rotation, also known as the Sagnac effect. The angular velocity of the Earth is 72.921×10^{-6} rad s^{-1}. The velocity of the clocks relative to the surface of the Earth v and v_{E} is the velocity component in the east direction. The difference in the gravitational potential between the geoid and the instantaneous clock position is given by $g(\varphi)h$, where $g(\varphi) \approx 9.7803 + 0.0519 \sin\varphi$ m s^{-2} and φ the geographical latitude. The radius of

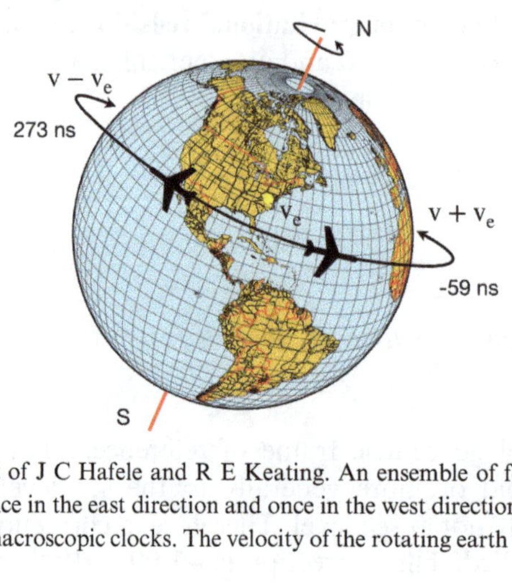

Figure 2.43. The experiment of J C Hafele and R E Keating. An ensemble of four cesium clocks was flown around the rotating earth, once in the east direction and once in the west direction, to test the predictions from the theories of relativity on macroscopic clocks. The velocity of the rotating earth was v_{e} and the velocity of the plane was v.

the Earth is a_1 and finally we have τ, the display of the traveling clock and t the respective display at sea level. The result from the experiment was a clock correction of 59 ± 10 ns for the flight going east and 273 ± 7 ns for the journey going west. The error margin is larger than the uncertainties of the clocks. This is caused by accumulated uncertainties of the integrated flight path. This experiment illustrates in an impressive way how important the GNSS orbit determinations for gravity missions like CHAMP, GRACE and GOCE are.

In the eighties of last century, before the times of GNSS, the clocks between Wettzell and Braunschweig were compared side by side by a traveling clock. For that purpose there was a portable rubidium clock available. This clock was first compared to the Cs master clock in Wettzell and after traveling to Braunschweig it was compared to the reference clock at the PTB. The travel velocity during the transport did not exceed 80 km h^{-1}. In the end the required clock correction from the transport velocity turned out to be -1.2 ns.

2.5.1 A simple clock error model

We reiterate, clocks are defined by the origin of their time scale, the clock rate and the direction of counting. For an ideal clock the clock rate, in other words the scale is constant and we can write[6]:

$$T_{\mathrm{I}} = \frac{1}{f_{\mathrm{I}}} \tag{2.45}$$

A time interval $(t - t_0)$ can be determined by counting n_{I} oscillations.

$$t - t_0 = n_{\mathrm{I}} T_{\mathrm{I}} = \frac{n_{\mathrm{I}}}{f_{\mathrm{I}}} \tag{2.46}$$

In an alternative presentation we can express n_{I} as the integral of the clock frequency over the time interval reaching from t_0 to t:

$$n_{\mathrm{I}} = \int_{t_0}^{t} f_{\mathrm{I}} \,\mathrm{d}t = f_{\mathrm{I}}(t - t_0) \tag{2.47}$$

The quantity n_{I} corresponds therefore to the number of oscillations, which have elapsed since the start at t_0. Real-world clocks do not have a strictly constant frequency. However, we can approximate the frequency of a real-world clock f_{r} with a straightforward error model.

$$f_r = f_{\mathrm{I}} + \Delta f + \dot{f}(t - t_0) + \tilde{f}(t) \tag{2.48}$$

In this expression Δf describes a constant offset from the ideal frequency (error in scale), $\dot{f}$ takes care of a drift in frequency and $\tilde{f}$ stands for arbitrary frequency errors (noise). Inserted into equation (2.48) we obtain a count of the oscillations of a real-world clock:

---

[6] The index I stands for ideal.

$$n_{\mathrm{r}} = \int_{t_0}^{t} f_{\mathrm{r}}(t)\mathrm{d}t = f_{\mathrm{I}}(t - t_0) + \Delta f(t - t_0) + \frac{\dot{f}(t - t_0)^2}{2} + \int_{t_0}^{t} \tilde{f}(t)\mathrm{d}t \quad (2.49)$$

When we convert this expression to show the determined time interval, we get:

$$(t_{\mathrm{r}} - t_0) = n_{\mathrm{r}} T_{\mathrm{I}}(t - t_0) + \frac{\Delta f}{f_{\mathrm{I}}}(t - t_0) + \frac{\dot{f}(t - t_0)^2}{2f_{\mathrm{I}}} + \int_{t_0}^{t} \frac{\tilde{f}(t)}{f_{\mathrm{I}}}\mathrm{d}t \quad (2.50)$$

Now, we also assume that there was a synchronization error of the real-world clock for the start epoch of the time interval. We get:

$$\Delta t_{\mathrm{r}}(t_0) = n_0 T_{\mathrm{I}} \quad (2.51)$$

We can establish the departure of the real-world clock from the ideal clock by

$$\Delta t_{\mathrm{r}}(t) = t_r - t = \Delta t_r(t_0) + \frac{\Delta f}{f_{\mathrm{I}}}(t - t_0) + \frac{\dot{f}(t - t_0)^2}{2f_{\mathrm{I}}} + \int_{t_0}^{t} \frac{\tilde{f}(t)}{f_{\mathrm{I}}}\mathrm{d}t. \quad (2.52)$$

To make the expression clearer laid out, we rename some of the terms and get

$$t_{\mathrm{r}} = T(t_0) + R(t - t_0) + \frac{D}{2}(t - t_0)^2 + \int_{t_0}^{t} y(t)\mathrm{d}t \quad (2.53)$$

In this error model we find the constant offset of the real-world clock in comparison to an ideal clock in the first term $T(t_0)$. The second term $R(t - t_0)$ expresses the drift of the clock and the quadratic term $\frac{D}{2}(t - t_0)^2$ captures the aging of the clock, which shows up as a changing drift rate over time. The final term $y(t)$ includes the stochastic errors in the model. The application of clocks in space geodesy is based on real-world clocks and we can conclude that under normal conditions we can approximate the clock behavior with a quadratic polynomial to sufficient accuracy.

2.6 Delay compensation for optical time and frequency transfer

For the time transfer we need to establish the transport time for the clock comparison as well as the delay this signal transfer operation is subjected to. These delays arise from the signal propagation inside the instrumentation that facilitates the comparison.

2.6.1 Optical frequency transfer

When light pulses or electrical pulses travel between a source and a detector, either in free space or in a transmission line (cable or fiber), they take to cover the distance. They also take time to get recorded. When we attempt to transfer time, all these delays have to be detected and removed from our measurements. Temperature variations, bending or stretching of wires or fibers make these delays variable over time. Therefore, it is not enough to just calibrate these delays and subtract them from the measurements. A better approach is constantly measuring the delays or to control the signal path continuously. A prominent application is the frequency comparison of two remote optical clocks over a fiber link. This has been achieved over a distance of more than 1400 km, between Braunschweig and Paris [11].

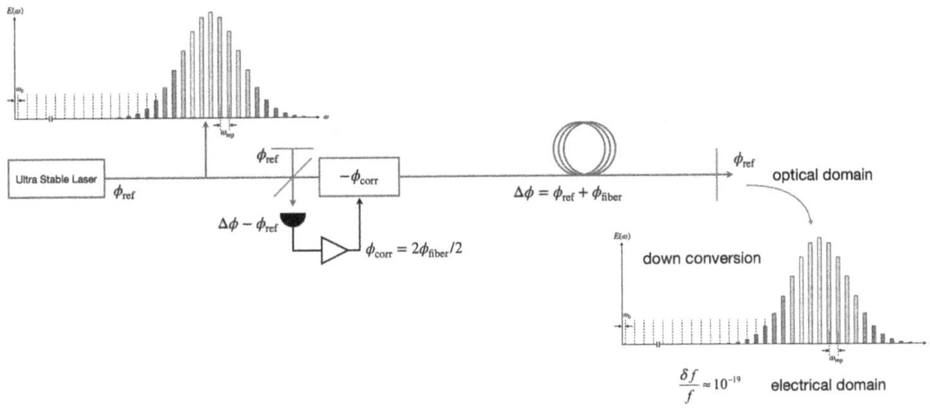

Figure 2.44. Illustration of the active delay control for a fiber link for optical clock comparisons.

Temperature and strain-induced refractive index variations are significant on a fiber line of that length. This causes considerable phase fluctuations, requiring active control. This active delay compensation is realized in the following way and illustrated in figure 2.44. Let us assume that we want to transfer the optical frequency of an ultrastable clock laser over a large distance of say nearly 1000 km [3]. In the absence of the active control system we would experience a considerable jitter of the phase. Since frequency is the derivative of the phase, this would amount to frequency variations over a large spectral range. The essence of the delay compensation is a two-way transfer setup. In other words, we transmit the optical frequency through the fiber and put a mirror at the other end, so that most of the light is back-reflected to the source. Here, we interfere the outgoing light with the reflected light that has picked up all the variable delay during its passage through the fiber. A closer look at figure 2.44 reveals that this setup is essentially a Michelson Interferometer with unequal arm length. The interferogram is detected by a photodetector and the output signal contains the phase fluctuations in the electrical domain. This signal is divided by two, since it contains the phase fluctuations of a double transfer through the fiber. Then it is inverted and added to the outgoing optical signal, utilizing an AOM. The result is an unperturbed frequency output at the end of the transmission line because any perturbation picked up on the way is instantly subtracted from the signal in this interferometer setup. The achieved frequency stability of these transmission lines was demonstrated to be around $\delta f/f \approx 10^{-19}$, which is about one order of magnitude better than the stability of well performing optical clocks. One might wonder why we use optical frequencies for the transfer and no microwave signals. As indicated in figure 2.44, the optical frequencies require a fairly involved optical frequency comb setup to actually make use of the transferred frequencies. However, the two-way approach that we exploit here is not available in the microwave regime. With electronic frequency signals we cannot realize this interferometer setup and consequently, there would be no compensation signal available to us.

2.6.2 Optical time transfer

The concept presented in the previous section transfers frequencies with no degradation in quality over long distances. In this section we look at the delay compensation for the transfer of time. Transferring time automatically means that we are dealing with high-bandwidth signals. For that matter we are using a mode-locked fs-pulse laser. These lasers have excellent low- noise properties, both in the optical and the microwave domain. We use the stable low-noise 100 MHz frequency of a hydrogen maser as a source. With that we lock the repetition rate of the mode-locked laser. As a result we obtain a continuous pulse train with pulses as sharp as 10 fs spaced apart by exactly 10 ns. This pulse train can be regarded as an optical ruler, against which we can derive signal propagation delays. In order to stabilize these delays, we again use an unequal arm Michelson Interferometer approach [16], which is graphically sketched in figure 2.45. The pulse train from the mode-locked laser passes a beamsplitter, where most of the light is guided to the long arm of the interferometer. This arm can be several hundred meters long. Furthermore, half of the length of the fiber run must be composed of a polarization- maintaining fiber and the other half composed of a conjugate fiber. This ensures that the dispersion, which spreads out the sharp pulses, is compensated. At the end of the fiber we still have the original sharp pulses available. A mirror at the end of the fiber line reflects most of the pulses back, also realizing a two-way transfer system. When the pulses have come back to the beamsplitter, the two arms are superimposed in a cross-correlator. A closed-loop feedback system drives a delay line in the long arm such that the outgoing and the reflected pulses stay on top of each other. The pulses that are transmitted at the end of the long arm are still equidistant in time with a 10 ns spacing and the fs-pulses are providing stable markers

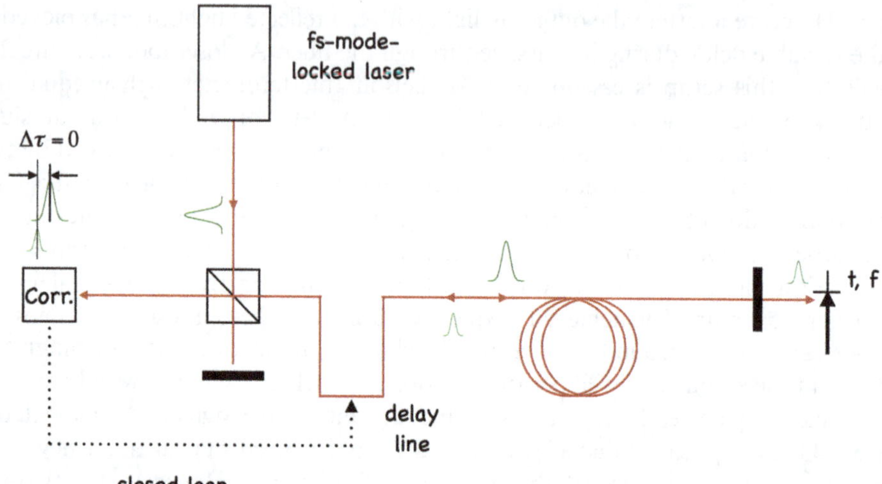

Figure 2.45. Illustration of the active delay control in a fiber link for the simultaneous transfer of time and frequency in an unequal arm Michelson interferometer configuration. The system exploits a mode-locked fs-pulse laser. For better readability only one pulse of the pulse train is shown in the various sections. Reprinted from [16]. CC BY 4.0.

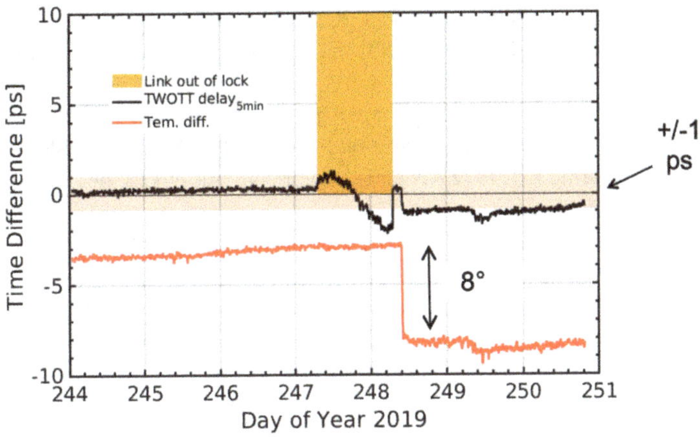

Figure 2.46. The variation of the time delay between two radio telescopes is presented as the black line and stays well within 1 ps. When the system is put out of lock, the stability is lost. The red line shows the temperature in the maser room. When it dropped by 8° due to a failure of the air-conditioning, the active delay compensation made up for this. Image credit: Jan Kodet.

for time. The observed stability for these markers between the signal source and the end of the line is $\Delta t \leqslant 1$ ps [10]. In order to provide an ideal common clock to a geodetic fundamental station, which collocates several observing systems in one spot with a stable uniform timebase, not just one of these delay compensated fiber lines, but at least one line per measurement system is required, except for VLBI, which needs two lines each, one at the feedhorn and one at the digitizer. Therefore, the full delay compensated time and frequency distribution system is arranged in a star configuration, where one source is split up to allow up to 14 individual compensation lines.

To demonstrate the capability of the delay compensation system, we discuss an example of a measurement series in figure 2.46. The time delay between two radio telescope antennas is compared over several days with an independent two-way optical time transfer system (TWOTT) and the relative delay between the two feed-horns is shown as the black line in the diagram. On day four the delay compensation was deliberately taken out of lock for one day. One can see that the relative delay starts to drift between the two systems. When the system was put into lock again, it came back to its original delay setting. A couple of hours later, we had a failure of the air conditioning in the maser room and the temperature dropped by 8°. We can see that the delay compensation for time and frequency was only marginally affected. It almost remained in the intended stability window of ±1 ps. We note a change of delay of 3 ps for optical ranging corresponds to 1 mm in distance.

References

[1] The time scales TAI and EAL www.ptb.de

[2] Brückner F, Friedrich D, Clausnitzer T, Britzger M, Burmeister O, Danzmann K, Kley E -B, Tünnermann A and Schnabel R 2009 Realization of a monolithic high-reflectivity cavity mirror from a single silicon crystal *Phys. Rev. Lett.* **104** 163903

[3] Droste S, Ozimek F, Udem T, Predehl K, Hänsch T W, Schnatz H, Grosche G and Holzwarth R 2013 Optical-frequency transfer over a single-span 1840 km fiber link *Phys. Rev. Lett.* **111** 110801

[4] Einstein A 1973 *Über die spezielle und die allgemeine Relativitätstheorie* (Vieweg) pp 14–44

[5] Feynman R 1970 *The Feynman Lectures on Physics* **vol 1** (Addison-Wesley) pp 15-2–8

[6] Hafele J C and Keating R E 1972 Around-the-world atomic clocks: observed relativistic time gains *Science* **177** 168–70

[7] Hafele J C and Keating R E 1972 Around-the-world atomic clocks: predicted relativistic time gains *Science* **177** 166–8

[8] Heavner T P, Donley E A, Levi F, Costanzo G, Parker T E, Shirley J H, Ashby N, Barlow S and Jefferts S R 2014 First accuracy evaluation of NIST-F2 *Metrologia* **51** 174–82

[9] Hänsch T W 2006 Nobel lecture: passion for precision *Rev. Mod. Phys.* **78** 1297–309

[10] Kodet J, Schreiber K U, Eckl J, Plötz C, Mähler S, Schüler T, Klügel T and Riepl S 2018 Co-location of space geodetic techniques carried out at the Geodetic Observatory Wettzell using a closure in time and a multi-technique reference target *J. Geod.* **121** 6109–18

[11] Lisdat C *et al* 2016 A clock network for geodesy and fundamental science *Nat. Commun.* **7** 1215–23

[12] Ludlow A D, Boyd M M, Ye J, Peik E and Schmidt P O 2015 Optical atomic clocks *Rev. Mod. Phys.* **87** 637–701

[13] McCarthy D D and Seidelmann P K 2009 *Time: From Earth Rotation to Atomic Physics* (Wiley)

[14] Panfilo G and Arias F 2019 The coordinated universal time (UTC) *Metrologia* **56** 042001

[15] Ramsey N F 1990 Experiments with separated oscillatory fields and hydrogen masers *Rev. Mod. Phys.* **62** 541–52

[16] Schreiber K U and Kodet J 2017 The application of coherent local time for optical time transfer and the quantification of systematic errors in satellite laser ranging *Space Sci. Rev.* **214** 1371

[17] Steel W H 1986 *Interferometry* (Cambridge University Press)

[18] Sullivan D B, Allan D W, Howe D A and Walls F L 2001 Characterization of clocks and oscillators *NIST Technical Note* **1337** pp 1–357

[19] Takamoto M, Ushijima I, Ohmae N, Yahagi T, Kokado K, Shinkai H and Katori H 2020 Test of general relativity by a pair of transportable optical lattice clocks *Nat. Photon.* **6** 1–7

IOP Publishing

The Measurement Techniques of Space Geodesy
Optical technologies
Karl Ulrich Schreiber

Chapter 3

Satellite laser ranging (SLR)

Satellite laser ranging (SLR) is a very unique technique in the toolbox of space geodesy. It stands out because it operates in the optical domain. Furthermore, it is the only technique that operates in a two-way configuration, where the laser signal passes through the measurement distance twice. Although this causes a significant limitation for the maximum distance that this technique can cover, it has the great advantage that unlike all the other measurement techniques of space geodesy, we only work with a single clock. The main goal of SLR is the accurate determination of satellite orbits. In order to achieve this, the path of the satellite is sampled as it moves across the sky of the observatory. Depending on the height of the orbit, each ranging station can only track a segment of the complete orbit. From high orbits we get larger segments, from low orbits we get smaller segments. There are two ways to reconstruct the satellite orbit. One possibility is the accurate determination of the pointing direction (angle) as a function of time, the other is the measurement of the distance (range). In order to have a useful contribution to the geodetic reference frames, a measurement resolution of $\Delta x/x \leqslant 10^{-9}$ is required. This practically excludes angle measurement concepts, as something of the order of $\delta\phi < 100\,\mu$as is exceedingly difficult to obtain. Since distance and time can both be related to the vacuum speed of light, the range measurement is a promising goal. We recall that time is particular among the SI units, since it can be established with the highest resolution of all. The orbit of the LAser GEOdynamics Satellite has a radius of approximately 12 800 km and a desired measurement resolution of $\approx$1.2 cm is a challenging but reasonable goal. Laser ranging is performed by about 30 SLR stations around the world. They are organized and coordinated by the *International Laser Ranging Service (ILRS)*, an entity under the roof of the *International Association of Geodesy (IAG)*. A specific introduction to SLR, the network and the involved procedures for the operations can be found in [2]. In the following we are dealing with the technical aspects of SLR.

doi:10.1088/978-0-7503-2032-0ch3

3.1 Overview

Laser ranging is unique among the measurement techniques in space geodesy. It is based on the measurement of the time of flight of short laser pulses [5]. The measurement process itself is carried out entirely in the optical regime. Furthermore, it is a two-way technique. That means that there is only one clock involved. We measure accurate intervals of time and not specific epochs, therefore a clock offset does not matter. Unlike the other techniques, GNSS or VLBI, which involve several independent clocks, there is no need for a clock synchronization. The result of SLR is directly obtained, along with true distances between a reference point on the ground and a reference point on the satellite. Laser pulses in a vacuum travel at the speed of light c_0, and this is a universal constant derived directly from the Maxwell equations.

$$c_0 = \frac{1}{\sqrt{\varepsilon_0 \mu_0}} = 299\ 792\ 458\ \mathrm{m\ s^{-1}} \tag{3.1}$$

Another advantage is the well-defined path of propagation. Both origin and endpoint of the range measurement are well defined and there are no multipath propagation effects present. Figure 3.1 illustrates the concept of SLR in the form of a block diagram. A pulse laser generates short mode-locked laser pulses with a typical pulse width of 10 ps $\leqslant \tau \leqslant$ 200 ps. These pulses are transmitted by a tracking telescope toward a suitable satellite. When the satellite is illuminated by the laser pulses, a small fraction of the light is reflected back toward the ground station with the help of cube-corner reflectors (CCRs). The returning light can then be detected by a sensitive broadband photonic detector. By obtaining the roundtrip time of the laser pulse, the distance between the ground station and the CCR at the epoch of the laser fire event can be established.

$$r(t_0) = \frac{1}{2} c \Delta t \tag{3.2}$$

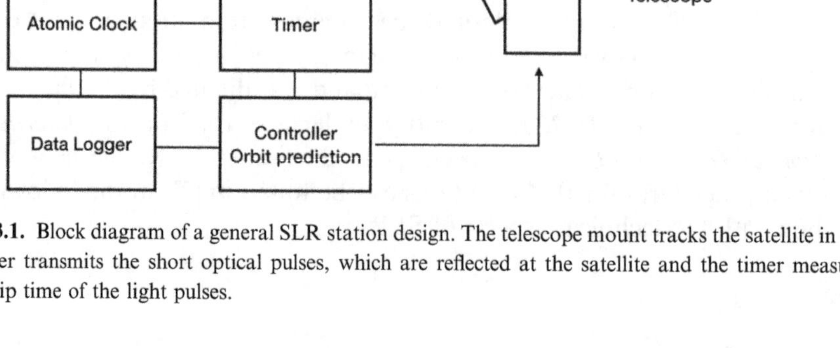

Figure 3.1. Block diagram of a general SLR station design. The telescope mount tracks the satellite in the sky. The laser transmits the short optical pulses, which are reflected at the satellite and the timer measures the roundtrip time of the light pulses.

Although this measurement principle is extremely simple, a highly productive SLR station is a very complex measurement device. This is caused by the high measurement resolution and the stringent demands on measurement stability. SLR can be carried out during the daytime and at night, so a very small field of view and excellent tracking properties with an uncertainty of about 10 seconds of arc or less are required. This in turn places high demands on the telescope control hardware and the *a priori* knowledge of the satellite orbit, so that a sufficiently accurate telescope pointing can be facilitated. This requires a linear scale and an encoder resolution at the second of arc level.

The laser pulse is not time tagged at the reference point of the system, where the roundtrip time of the probe signal begins. Therefore, a postprocessing procedure has to correct for this kind of optical system delays. Since the intensity I of the laser pulse as a function of distance on the free space path falls off with $I \approx 1/r^4$, only a tiny amount of light is returning. The dynamic range between the outgoing laser pulse and the returning echo can be as large as 10 orders of magnitude with 2 or 3 orders of magnitude of variation during the tracking procedure. Therefore, a number of efficient filter methods are required to extract the returning light pulse of the laser from the overwhelming level of background radiation. Another complication is the portion of the light path, where the signal goes through the atmosphere. While turbulent density fluctuations cause waveform distortions and variable refraction, also the velocity at which the signal travels is no longer the vacuum velocity of light. A multilayer correction model provides the necessary delay compensation. It uses the measured atmospheric pressure, taken at ground level near the reference point of the telescope, along with the temperature and the humidity.

Short laser pulses provide high resolution in time. Therefore, the pulse width (typical for Nd:YAG solid state lasers is 10–200 ps) is an important criterion. However, sufficient energy is also required in order to make up for the attenuation along the line of sight. Pulses shorter than 10 ps have not improved the measurement resolution. Wavefront distortions due to turbulent atmosphere have been found to stretch very short pulses to about 7 ps in time [6] and the limited bandwidth of the timing unit presents another bottleneck. This is followed by apparent target-depth effects for satellites, which have more than one contributing CCR installed.

Last but not least, there are also limitations in the resolution of the time interval measurement. This is caused by the limited bandwidth of electronic circuits, cabling, thermal drift of the hardware components and ground potential fluctuations in the trigger stages. Figure 3.2 illustrates how the measurement resolution has improved over time as better lasers and electronic components became available. A better range resolution of the ranging systems led to smaller errors in the orbit determination process and this made more and more geophysical processes quantitatively accessible. The horizontal lines in figure 3.2 indicate below which resolution limit the gravity field (a), the Earth rotation and tides (b), the tectonic plate motion (c) and plate deformations (d) could be resolved.

The best type of satellites for SLR approximate a dot mass in a mathematical sense. Ideally, they are small, spherical, heavy and uniform in their mass distribution. The

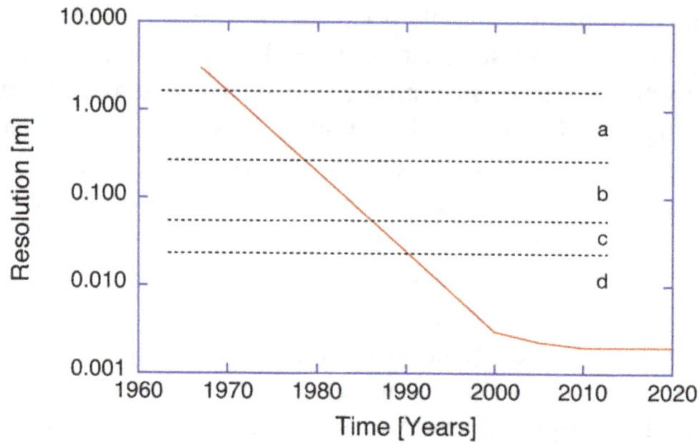

Figure 3.2. The development of the SLR range resolution over the years. The regime (a) marks the resolution limit, where the gravity field started to become accessible. Earth rotation and tides became visible in (b). Plate tectonics (c) and plate deformations (d) could be resolved at a range resolution of several cm.

LAGEOS
diameter 60 cm
mass 450 kg
≈ 6000 km above ground

STARLETTE
diameter 24 cm
mass 47 kg
≈ 900 km above ground

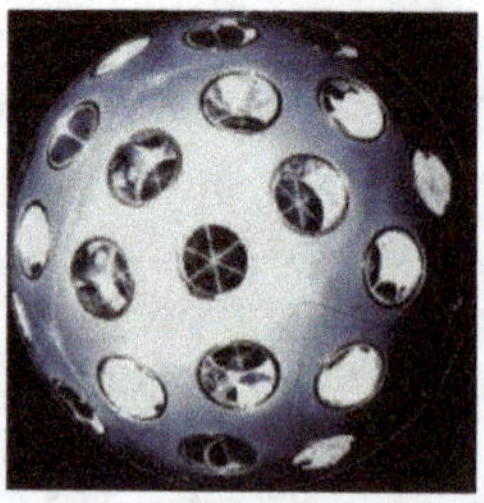

Figure 3.3. The satellites LAGEOS and STARLETTE are good examples of geodetic 'cannonball' type of satellites. They were the first of their kind, launched in 1976 and 1975, respectively. Photo credit: NASA.

CCRs are placed uniformly all over the surface such that the offset between the center of mass and the effective reflection point of each reflector is the same. Figure 3.3 shows two examples of different sizes put into a high and a low orbit. STARLETTE has a diameter of only 24 cm in order to reduce orbit perturbations caused by friction from the residual atmosphere at a height of about 900 km above ground. LAGEOS needs to be larger in order to provide sufficient reflector cross-section, to compensate for the much larger distance. We note that the space segment of SLR satellites is entirely passive and technically simple. Geodetic satellites are long lasting and suitable for the construction of a stable frame of reference. The full technical complexity of SLR is located at the ground station. Therefore, the instrumentation is accessible and can readily be modified or modernized. The accessibility, however, comes with the complication that even small changes in the hardware layout can cause variations in the systematic error budget, which may compromise the station stability. Finally, it must be appreciated that SLR is an optical technique and measurements are only possible under clear sky conditions.

3.2 The instrumentation of satellite laser ranging

Ranging by measuring the time of flight is not an unusual concept. In fact, this method is widely used. Bats are ranging by utilizing ultrasound. Since the velocity of sound inside the atmosphere is of the order of 330 m s^{-1} it is also a common practice to determine the approximate distance of a flash of lightning by counting the seconds between the occurrence of the optical flash and the arrival of the sound of the rumble of the thunder. The parking aid of modern cars apply the same basic concept to determine the distance between the front or the back of a moving car with respect to stationary obstacles at the roadside.

Airports use radars to establish the path of flight of approaching aircraft. They apply time-of-flight measurements of microwave pulses, which are scattered off the aircraft structure. Figure 3.4 reduces the block diagram of figure 3.1 to a minimum. There is the transmitter and the detector, both of which are connected to the stop clock to establish the time interval between the epoch of transmission and the epoch of detection. In space geodesy we range to satellite targets in order to determine their orbits, and we have very high demands on the measurement resolution. A remaining uncertainty of the order of a few millimeter is desirable in order to obtain a stable global geodetic frame of reference. Such a stringent requirement is technically very demanding and this is where the complexity of SLR systems comes from.

In order to obtain an accurate range to a satellite, we require a well-defined point of origin and an equally well-defined endpoint for the distance measurement. The point of origin must be geometrically sharply defined and readily accessible to a local survey, so that it can be connected to a local survey grid with high accuracy. Furthermore, it must be completely independent of the telescope pointing direction. Alt-azimuth telescope mounts are widely used in SLR for their convenience of operation. They can be rotated on two axes, one around a horizontal and the other around a vertical axis, in order to reach each point in the sky. Since the azimuth and elevation axes intersect each other, this intersection point defines the origin of the range measurement. It is often called the invariant point. It is not affected by the telescope motion (see figure 3.1). Of similar importance is the endpoint of the SLR measurement. Geodetic cannonball satellites (see figure 3.3) have a definitive center of mass, which defines the orbit. Our range measurement only goes up to the surface of the satellite, where the laser light reflectors are located. Each of the reflectors has to be constructed such that the range measurement can be associated with any point of the reflector surface with a maximum variation of about 1 mm. Furthermore, all reflectors on the surface of the spherical satellite have to have the same distance to

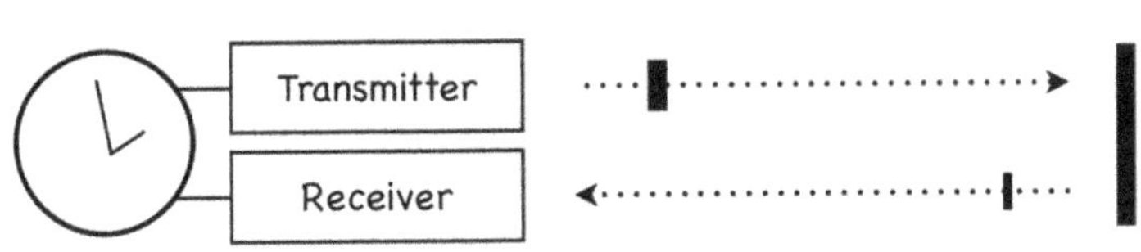

Figure 3.4. The concept of a time-of-flight measurement for ranging is a very general and widely used concept. Unique for laser ranging is the application of ultra-short laser pulses and specific reflectors to gain better control over the point of reflection.

the center of mass and this distance must remain constant and has to be known accurately as well.

The duration of the laser pulse determines the achievable resolution of the range measurement. The shorter our probe signal is, the sharper is the detector response in time and this eventually determines the obtained range resolution. Pulsed ruby lasers were the first transmitters in the early days of SLR. They offered a pulse width of 3–5 ns (or a spread of 0.9–1.5 m), a sensation at the time. Modern Nd:YAG lasers deliver pulses, 10 ps or a third of a cm long, but this is not the only difference. While the ruby laser could produce about one pulse per minute, the Nd:YAG lasers have a repetition rate of up to 2 kHz. Repeating a measurement N times improves the uncertainty of the range measurement by $1/\sqrt{N}$, provided that the measurement errors are normal distributed and free of systematics.

Finally, it is important how well the timer can tag an epoch of measurement, and we need to look at two aspects here. The most obvious necessary property is a very wide bandwidth of the clock electronics as well has an extremely small granularity or fine resolution of the measured time. We want to derive timing with a resolution not worse than a few ps, so this requires an excellent clock. While resolution is the most obvious property of the timer, the stability and accuracy of the time base of the clock is of similar importance. We measure the time of flight, therefore the clock oscillator has to represent the second as defined by the SI system, otherwise we inherit an error in scale. In addition to that, we also demand that there is no perceptible drift in the clock, which would also act detrimental to the desired reliability of the sampled satellite orbit. This effect is range dependent.

3.2.1 The laser system

The word LASER has been coined from a description of the underlying principle of operation. 'Light Amplification by Stimulated Emission of Radiation'. In the early days, lasers were often called optical masers, which underlines the close relationship between these two types of instruments. In both cases the operations are based on discrete excited atomic states inside a resonator, where the desired radiation caused by spontaneous emission is fed back in order to obtain amplification through stimulating further emission processes. The most significant difference between a maser and a laser is the oscillation frequency. A maser typically operates in the microwave regime, while lasers work in the optical domain. Two mirrors and a gain medium are required as a minimum to form a resonant optical cavity and laser oscillation as sketched in figure 3.5. Lasing is supported as long as the gain exceeds the losses. This requires a closed optical path on which the light can bounce back

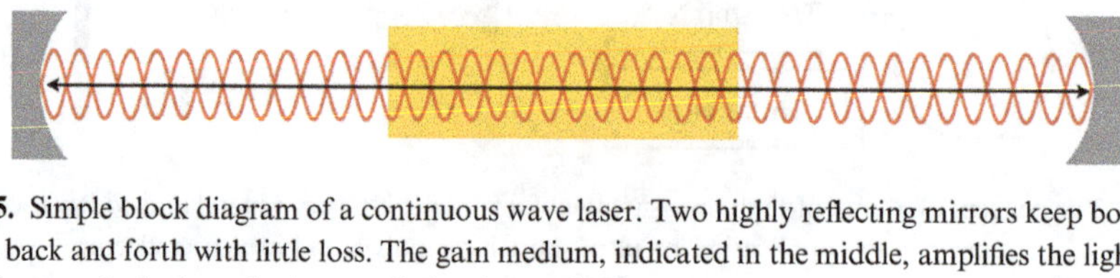

Figure 3.5. Simple block diagram of a continuous wave laser. Two highly reflecting mirrors keep bouncing the radiation back and forth with little loss. The gain medium, indicated in the middle, amplifies the light. As long as the gain exceeds the loss, the laser radiation is sustained.

and forth, similar to an electronic oscillator with a positive feedback amplifier attached. The gain medium amplifies the light and as long as the overall loss of a roundtrip of the laser light is smaller than the gain, the laser will continue to generate light. The exact operation frequency of the laser depends on two conditions. It must be within the bounds of the atomic transition and the length of the cavity L must be an integer multiple N of one half of the wavelength $L = \frac{N}{2}\lambda$, or there is no coherent amplification.

An atomic transition allows an electron to go from a discrete lower energy state to a higher state by absorbing a photon of matching energy $\Delta E = h\nu$. Likewise it may drop from an excited upper state to a lower energy state by emitting a photon of the same energy $\Delta E = h\nu$. In order to generate gain, there must be more atoms in the excited state, otherwise absorption and emission would cancel each other out. Inversion cannot be achieved in general on every atomic species, therefore lasers are confined to some discrete transitions only and even those may differ greatly with respect to the available gain. Most lasers involve four different atomic energy states with different lifetimes in order to achieve inversion and hence the ability to amplify light coherently. Figure 3.6 illustrates this process[1]. At the beginning, the atom is in the lower energy state or ground state, denoted by E_0. From here, it is excited into the upper state E_3 by an absorption or collision process (1). The upper state E_3, however, is short lived. This causes the electron to quickly relax into the long-lived energy state E_2, which at the same time is the upper state of the desired laser transition (2). Due to the fact that this state is long lived, the system reaches a state of inversion as many atoms accumulate electrons in this excited state. Stimulated emission (3) provoked by the spontaneous emission of light circulating inside the laser cavity causes the electrons to

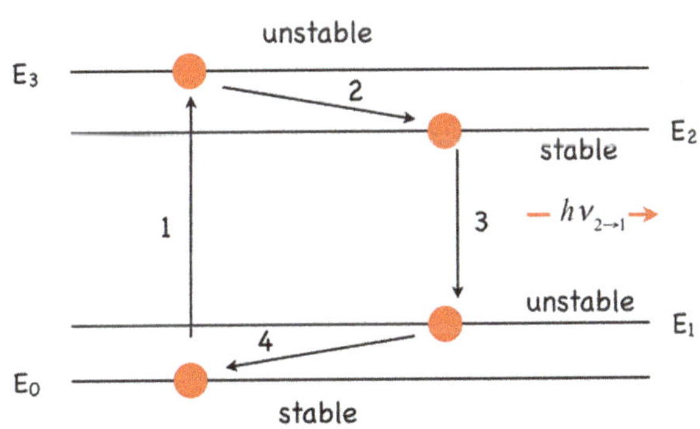

Figure 3.6. Graphical illustration of the laser amplification process. Through an absorption or collision process the electron is lifted from the ground state E_0 to the short-lived upper state E_3. It relaxes quickly into the long-lived upper laser state E_2, from which it decays into the short-lived lower laser state E_1, either by spontaneous or stimulated emission. Inversion is maintained because the electrons quickly relax into the original ground state E_0, leaving for the majority of atoms always much more electrons in E_2 than in E_1.

[1] Suggested reading: Dieter Meschede: *Optics, Light and Lasers* (ISBN 978-3-527-41331-7), Anthony Siegman: *Lasers* (ISBN 0-935702-11-3) or similar.

decay into the lower laser transition E_1, which again is rather short lived. Therefore the electrons quickly relax back into the stable initial ground state E_0, where the whole process started (4). Continuous pumping of the gain medium restarts this cycle over and over again. The efficiency of the pumping process (1) eventually determines the gain that the exploited laser transition exhibits.

Until now we have looked at continuous wave laser excitation only. The result of this process is a continuous flow of phase coherent photons and it can be visualized like a continuous sine wave at a wavelength that is compatible with the geometrical size of the resonator and the frequency band of the laser transition. Such a probe signal is unsuitable for laser ranging, since we do not have access to a particular phase angle of this wave. In order to perform the time-of-flight measurement, we need short pulses with a pulse width of the order of 10–100 ps, rather than a continuous wave train. Mechanically chopping the light, no matter how fast this can be performed, is not the way to achieve this. Short pulses can be obtained by an interference process, which is usually referred as a **mode-locking** process.

In order to obtain light amplification through stimulated emission radiation, the incoming stimulating photon has to be of the same wavelength and in phase with the newly radiated photon. The feedback introduced by the second mirror only allows amplification for a wavelength of $L = \frac{N}{2}\lambda$. For a typical SLR system the number of wavelengths N in the laser cavity is of the order of 10^6. Provided the width of the laser transition is wide enough, we can excite quite a large number of different laser frequencies, all of which fulfill the lasing condition, however, for widely different values of N. Figure 3.7 shows the intensity distribution of about 20 different neighboring laser modes in the cavity. Without any precautions they are oscillating with a random phase relationship. This results in a highly variable intensity distribution. The whole situation changes when the phase relationship between the different laser modes is locked together. Figure 3.8 illustrates this by imaging a superposition of three different sine waves, which are tied together. At the places where all the amplitudes peak we accumulate light; in all other places, destructive interference is reducing the light level.

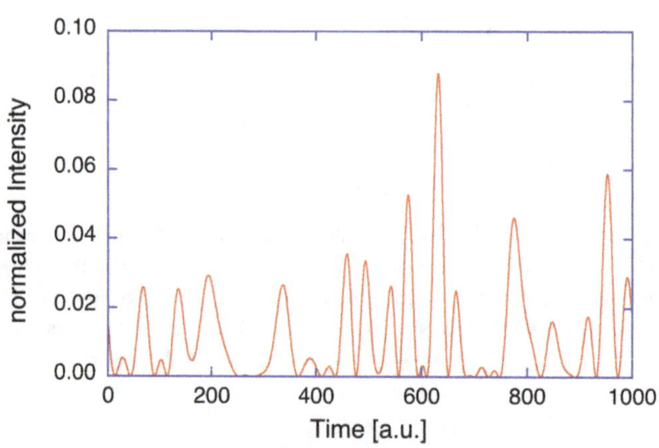

Figure 3.7. Superposition of 20 different optical frequencies with random phase relationship.

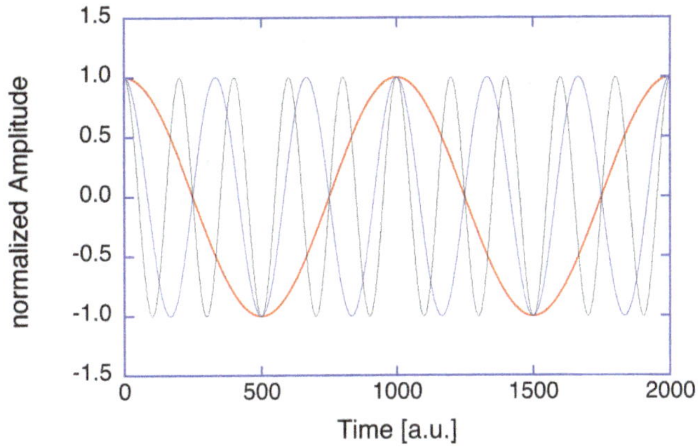

Figure 3.8. Superposition of three optical frequencies with fixed phase relationship (mode-locking).

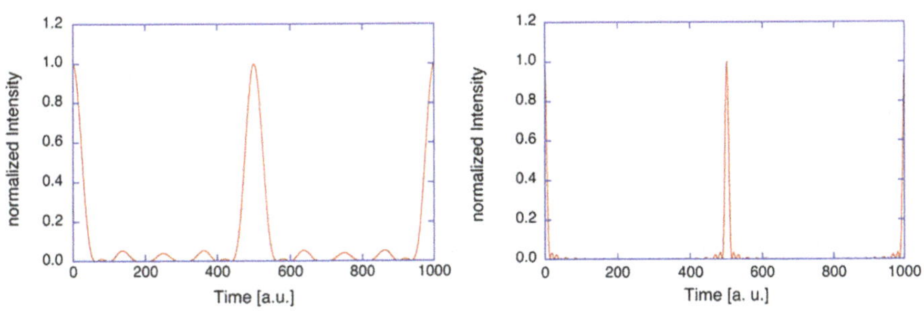

Figure 3.9. Illustration of how mode-locking leads to shorter and shorter pulses of laser light. In this example there are four different frequencies locked together on the left and twenty on the right.

As more frequencies are added, we obtain shorter and shorter bursts of high light level or in other words, shorter and shorter pulses of laser light. Figure 3.9 illustrates this process by showing an example of the output intensities generated when 4 or 20 frequencies are mode-locked together. The width of the exploited laser transition determines the ultimate pulse width that one can obtain from a given gain medium. In the early days of SLR, ruby lasers were in fashion. They eventually provided a pulse width of the order of 3 ns, which means that the sounding laser pulse was about 1 m long. For the roundtrip measurement this corresponds to a range resolution of 0.5 m. The laser transition in Nd: YAG (neodymium yttrium aluminum garnet) at 1.064 nm is wider ($\approx$120 GHz) and allows laser pulses as short as 10 ps or 3 mm in length.

Mode-locking is one of the necessary ingredients for the generation of short high-power laser pulses that can be used to range to remote satellites. It provides the short pulses; the other important feature is the Q-switching of the laser cavity in order to obtain sufficient energy in each pulse. The energy content or the number of photons present in the laser pulse is governed by the achieved inversion in the gain medium and corresponds to the number of electrons in the energy state E_2 (see figure 3.6). The Q switch acts like a digital switch, which makes the laser cavity lossy while the upper laser state is pumped to increase the level of inversion and then within a few ns it becomes transparent, so that high-energy pulses can quickly build up. While saturable absorbers (bleaching dye in solvents) have been frequently used in the past, they have made way for acousto-optic modulators and polarization switching Pockels cells [9, 11]. Pulses with an energy of 100–500 μJ are derived from such laser oscillators at a repetition rate between 5 and 20 Hz, which is not enough for the ranging of high-altitude satellites. On closer inspection we find that there is more than one single laser pulse generated. Every time when the laser pulse bounces back and forth inside the cavity some of the light leaks out at the front mirror. For as long as the inversion of the gain medium is not exhausted we get one pulse per roundtrip after the next, typically 10–20 for each shot. Therefore we have to apply a pulse picker, typically two concatenated Pockels cells, exercising fast polarization switching in order to allow the single pulse with the highest intensity to pass, while all other pulses are dumped out of the beam path. Nd:YAG laser pulse amplifiers are therefore used to bring the pulse energy up to the required level. Figure 3.10 illustrates the design of such a high energy pulse laser schematically. Practical systems have between one and three of these amplifiers, covering an energy range between 5 and 100 mJ. Such amplifiers differ from laser oscillators in that they do not have mirrors forming a cavity, so that they can provide the necessary feedback for starting the laser oscillation. While the gain medium in the amplifiers are pumped in the same way and at the same time as the oscillators, they only offer a single-pass transfer for the laser pulse to gain energy from the inversion inside the solid state laser crystal. Usually, each of these single-pass amplifiers increases the energy of the laser pulse by a factor of 5–10, depending on the level of saturation of the gain medium. Nd:YAG lasers operate in the near infrared regime at a wavelength of 1.064 μm and this requires a few compromises. Until about the year 2015 there were no suitable photodetectors with a high quantum efficiency and fast signal response for the infrared available. Therefore, all the SLR systems applied a crystal with nonlinear properties for frequency doubling. The output signals were laser pulses with $\lambda = 532$ nm, which corresponds to the regime of the green color. For this wavelength quite a variety of different fast rise time photodetectors with quantum efficiencies of $12\% \leqslant QE \leqslant 70\%$ existed, so that the loss of laser power due to the conversion process was more than compensated. Silicon based photodetectors in the infrared regime typically offered a quantum efficiency of up to 0.5%–4% at reasonably low intrinsic detector noise. The alternative InGaAs detectors were about 5 times more efficient, but exhibited a prohibitively high dark count rate of several MHz

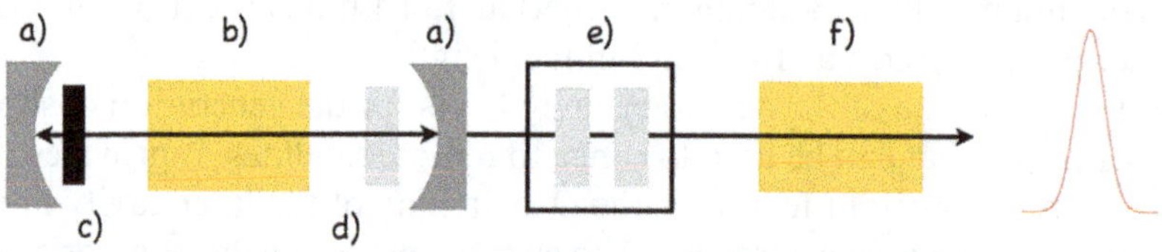

Figure 3.10. Schematic of a ranging pulse laser. The oscillation builds up between the mirrors (a) from the inversion in the gain medium (b). Q switching (c) and mode-locking (d) generate a pulse train of 10–20 pulses. Polarization switching with 2 Pockels cells (e) picks a single pulse out of the train, which is subsequently amplified in a single-pass amplifier (f).

and the photosensitive area is too small to image the entire telescope aperture onto the detector. In section 3.2.3 we will look at this in more detail.

Between 2010 and now, there are more and more laser ranging systems that operate on significantly higher repetition rates in the regime of 1–2 kHz. Since one cannot squeeze more energy out of these systems, the ranging beam power is lower by the proportion of the repetition rates and turns out to be of the order of 1–2 mJ. A direct consequence is the operation in the single photon detection regime with special requirements for the signal detection process. We will discuss this mode of operation in more detail in section 3.2.8.

3.2.2 The telescope

The telescope is required to direct the short laser pulses to the target and then subsequently to collect the reflected light and image it onto the photodetector. This means either that the laser (transmitter) and the detector (receiver) have to be attached to the telescope output port in an alternating sequence. Alternatively the entire telescope structure can be set up such that it comprises two subsystems—a small telescope to transmit the laser pulse and another dedicated large telescope for the detection process. The former setup is called monostatic, the latter bistatic. Both types of realizations have their individual set of advantages and disadvantages, which have to be balanced so that the eventually realized design depends on the main operational objective for each specific instrument. High-repetition rate laser operation clearly favors a bistatic setup in order to avoid the need for fast switching between the transmit and receive section. Systems that also care for Lunar Laser Ranging (LLR), will prefer a monostatic setup, as it is very difficult to maintain the co-linearity to within a few seconds of arc between the two telescope axes of a bistatic telescope layout.

The telescope connects the laser to the sky and it can be turned into any direction, so that every point of the sky can be accessed. Most telescopes are built as alt-azimuth systems. That means they have a rotary table around a vertical axis (the azimuth platform), on which the lens tube sits, which can be slewed around a horizontal axis, in order to reach all positions at different elevation angles. Figure 3.11 illustrates this

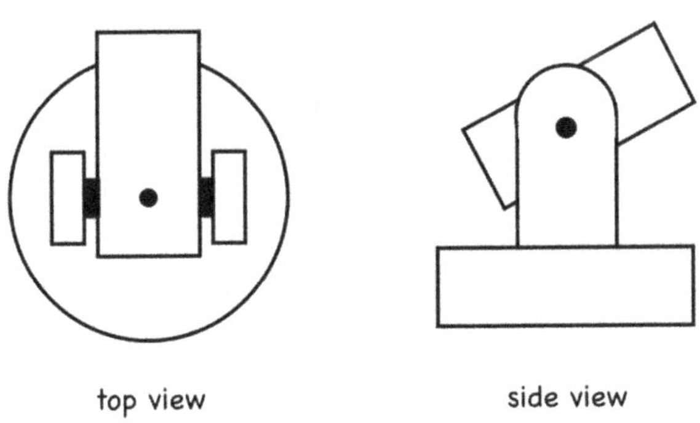

Figure 3.11. Basic illustration of a telescope in a monostatic design. The rotation around the vertical axis is illustrated when viewed from the top. The rotation around the elevation axis is shown when viewed from the side.

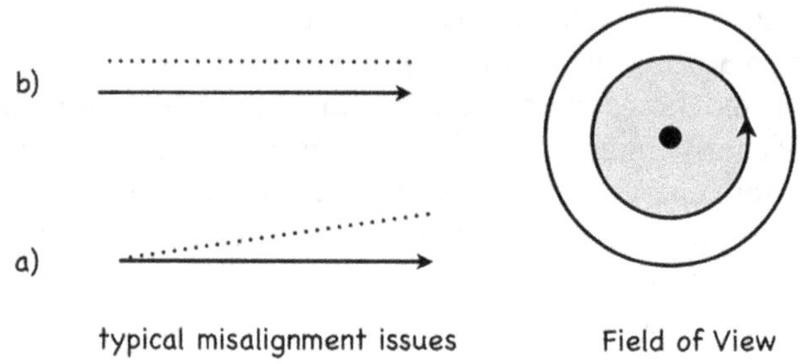

Figure 3.12. Graphical illustration of typical alignment issues on alt-azimuth telescopes. If the laser axis (dashed line) is parallel, but not on the telescope axis (b), the laser beam will rotate around the field of view as the telescope points around the sky. A considerable pointing error develops when the two axes are not collinear with respect to each other (a).

general design. The two axes of rotation have a common intersection point, which physically is in the middle of the lens tube. This point is geometrically the origin of our range measurement. Neither the laser output, nor the detector input are anywhere near this point. Therefore, it is required to establish the respective transmit and receive delays of the entire SLR system, in order to reference the time-of-flight measurement to this system reference point. The laser is usually located several meters away from the telescope. This means that the laser beam has to be guided exactly through the boresight of the vertical and horizontal axis in order to point correctly to the desired position. If the beam is offset from the boresight but still collinear, the laser beam will rotate around the optical axis of the telescope off center, which causes problems with the detector, but does not compromise the beam pointing. If the laser beam is at a small angle with the boresight, this alignment angle error will also rotate around the telescope axis with the telescope motion and cause offset pointing. This can make the tracking in particular for high-altitude satellites extremely difficult when the alignment error is significantly larger than the divergence of the laser beam. Figure 3.12 illustrates these scenarios graphically. A typical value for the divergence angle of a ranging system for high earth orbiting satellites is of the order of 2–5 seconds of arc. If the alignment error of the laser exceeds this value, there are no satellite echoes detectable since the laser pulse misses the satellite. Even worse because of the fact that the laser beam is routed through the two moving telescope axes, the offset angle will not be at a constant value, but move around the sky. That means that the required offset pointing angle changes constantly, which makes it hard for the observer to keep the satellite illuminated. A considerable loss of observations is the typical consequence from even small laser misalignments, so that this error source requires careful attention.

Apart from pointing the laser beam at the target and smoothly track the satellite as it moves across the sky, the telescope also serves the purpose to reduce the divergence of the laser beam. By illuminating a smaller patch in the sky, the already very high signal loss is reduced. We can express the diffraction limit of a telescope as

$$\Phi = 1.22\frac{\lambda}{D}. \tag{3.3}$$

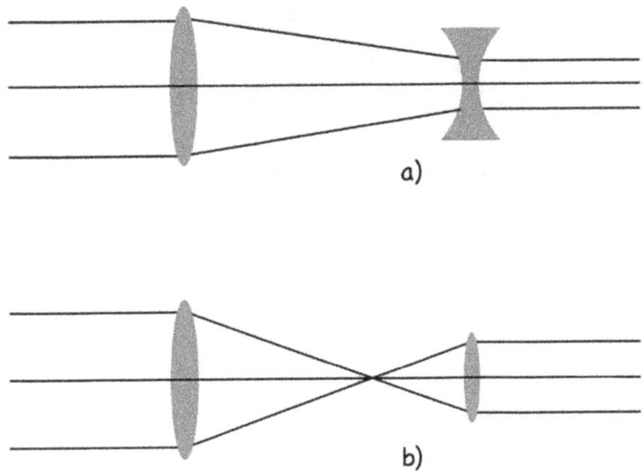

Figure 3.13. There are two types of telescopes. The Galilei telescope uses a convex and a concave lens (a) to enlarge the image resolution. This system is entirely afocal, which means that the image does not invert. The second concept is built from two convex lenses (b). Here we have a distinct focus and the image stands on its head.

Strictly speaking, this equation means that two objects viewed through the telescope with an aperture diameter D can be recognized as separate, when they are at least apart by the angle Φ at a given wavelength λ. It is obvious that the resolution of the telescope improves, the larger the aperture becomes. Applied to our laser ranging application, we find that the laser beam divergence for a 1 m telescope at the laser wavelength of 0.532 μm can be as low as 0.65 μrad, which is about 1/10 of an arc second. However, it would require a perfectly collimated laser beam at the telescope input port to achieve such a low beam divergence, which in reality is not there. However, for a lunar laser ranging station a divergence angle of the order of one second of arc is quite a typical operation parameter. Telescopes can have two very distinct imaging characteristics, commonly known as the Galilei and the Kepler configuration. They are illustrated in figure 3.13. The Galilei telescope is composed of a concave and a convex lens, which makes the telescope an afocal system. Historically this configuration has been widely used in seafaring because it only enlarges but does not flip the image, like the Kepler configuration does. This property of image inversion does not matter in astronomical applications, where the Kepler configuration is almost exclusively found in telescopes. Their advantage is the fact that the field of view can be conveniently adjusted by a variable aperture at the focus point.

In SLR we usually find both telescope configurations applied. The transmitted laser beam is passing through a Galilean telescope. The high energy density of the laser beam does not allow to have a focus in the beam path. This would cause significant losses from air ionization and plasma generation. The receiving telescope in contrast requires a very narrow field of view of the order of about 10 seconds of arc (50 μrad), so that most of the unwanted background light of the sky can be blocked off. Although figure 3.13 depicts a system that uses lenses, in practice most telescopes are actually mirror telescopes, which are cheaper to build and lower in loss.

Figure 3.14. Telescopes can be realized as monostatic or bistatic systems. An example of the former is the WLRS telescope at the Geodetic Observatory Wettzell, where outgoing and incoming light share the same path, while the latter concept features a separate transmit and receive path on two collinearly aligned telescopes side by side. Image credit: ZEISS Archive (left), SLR system at the Space Geodesy Facility Herstmonceux. Credit: Space Geodesy Facility Herstmonceux (right).

Unlike in astronomy, where earth rotation has to be compensated, most telescopes in SLR are alt-az mounts, since the satellite targets can follow any trajectory across the sky. Figure 3.14 illustrates the two different forms of telescope realizations that are used for ranging. In principle one can use the same telescope for the outgoing laser beam and the incoming echo returns. This, however, requires some switching element, somewhere along the Coudé path, which alternately connects the laser or the detector to the telescope. Due to the high transmit laser powers involved, this switching unit also involves the need for mechanical shutters, which finally limits the repetition rate of the laser to values of the order of about 500 Hz. Sharing the same telescope for both transmit and receive functions on the other hand allows a lower transmit laser beam divergence and usually provides a much better overlap of the outgoing and incoming beam. This makes such systems most suitable for long distance ranging, such as targets in the geostationary orbit and on the Moon.

Bistatic systems with their independent beam paths for the transmit and receive function do not have any constraints on the repetition rate on the measurement. They can also have their detector located very close to the telescope primary mirror, usually in the Cassegrain focus, which has the additional advantage that it reduces the receive beam path losses. However, the variable weight load of the two different telescopes as a function of the elevation pointing angle is often compromising the accurate collinearity of the transmit and receive telescope and therefore the overlap of the transmitter and detector field of view at the satellite target. This, however, is a practical technical but not a fundamental problem.

Mode-locked lasers with a pulse width between 10 and 20 ps delivering powers in the range of several watts, come as rather bulky and delicate instruments. They require water cooling and often contain high-voltage power supplies. That means that the

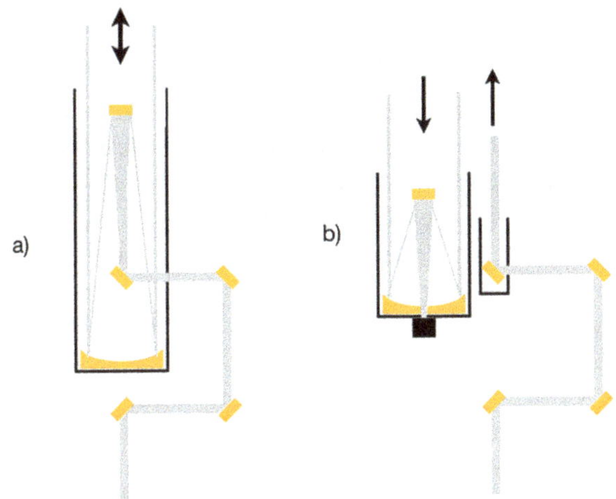

Figure 3.15. Illustration of the beam path in a SLR telescope for the monostatic case of a shared transmit and receive system (a) and the bistatic case, where there are two separate telescopes for the transmit and receive function (b). Telescopes with this design are shown in figure 3.14.

operation of a ranging pulse laser requires laboratory conditions, a stable optical table and sufficient space. Therefore, most of the SLR stations of the global network are feeding the laser signal into the telescope by a Coudé path. The advantage is a fixed position of the laser in the laboratory, while the telescope can point into any direction of the sky. However, this convenience comes with an additional difficulty. Since the telescope needs to rotate about its horizontal and vertical axis, the laser beam does not have a fixed alignment. It also rotates and gives rise to beam wander, if the laser beam path is not coaxial with the optical axis of the telescope. Figure 3.15 illustrates these two most common cases of a SLR system design. Although very rarely found, it is also possible to combine a small transmit telescope with a larger receiver aperture into a single telescope. By sharing the aperture between the transmit and receive function, a complicated mechanical transmit/receive switch can be avoided at the cost of sacrificing some of the captured return power, due to a blind spot on the receiver telescope, where the outgoing laser beam is injected into the telescope. Another interesting design feature has been implemented on the Apache Point lunar laser ranging system (Southern New Mexico), where the container of the 120 mJ ranging laser, including power supply and cooling is directly mounted behind the 3.5 m diameter primary mirror of the main telescope. This avoids any rotation of the field of view for the in particular very tight pointing requirements for LLR and has worked well over the years.

Typical operation parameters for transmit apertures within the ILRS are of the order of 5–10 cm. The receiving telescopes are usually between 50 cm and 1 m in diameter. The importance of these parameters are reflected in the link budget of the measurements and this is further discussed in section 3.4. Stations that range to the Moon require a lower beam divergence, and therefore are typically using the full aperture of the receive telescope. That means that these stations usually employ the monostatic system configuration.

3.2.3 Signal detection and timing

SLR measures the time of flight of a short laser pulse for a signal roundtrip between a ground station and a satellite target. The detection of the weak return signal makes use of the photoelectric effect[2] The incoming light is directed to the cathode of a photomultiplier tube, where the photoelectric effect releases one or several electrons. An applied electrical field between the cathode and a nearby dynode accelerates the electrons, so that they gain kinetic energy. When they hit the dynode, more electrons are released and the process starts again, causing a significant increase in electron numbers. Typical photomultipliers have 7–11 such dynode stages, before the cloud of electrons hits the anode, where a current is induced. Photomultipliers can have very high gain. They operate already at the single photon level. The basic concept is sketched out in figure 3.16. For the sake of clarity, the full voltage cascade for the dynode supply is not shown. The multiplication process takes place in a vacuum. The drawback of the photomultiplier is the limit in bandwidth. The multiplication process over several dynode stages broadens the cloud of electrons. This makes the laser pulse detection less sharp. Avalanche photodiodes (APD) operated in the Geiger mode are most widely in use for SLR today. They belong to the group of semiconductor components and the principle is sketched out in figure 3.17. The APD is operated in the reverse direction. Therefore, there is no current flowing across the diode when a voltage U is applied, as long as this voltage stays below the breakdown threshold. This bias voltage, creates a strong electric field in the junction between the n and p doped area. For high-bandwidth detection, the bias voltage is increased above the break-down threshold for a short period of say 1μs. In the absence of any free electron in the p-n junction region, an absorbed photon generates a free electron, which is accelerated in the strong electric field, creating an electron avalanche from collisions with other carriers, which then causes a breakdown and therefore a sharp rising edge at the

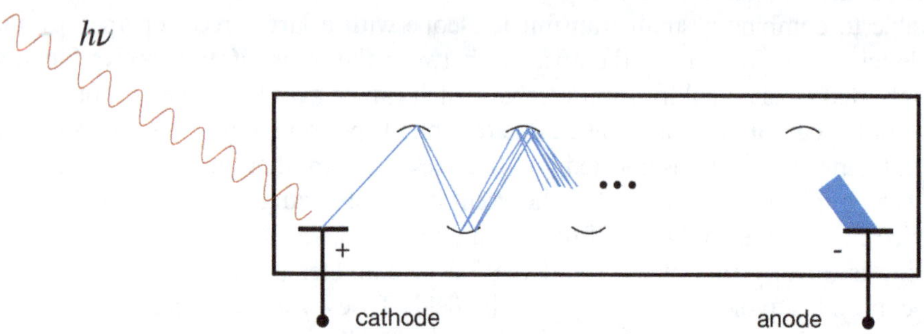

Figure 3.16. Operation principle of a photomultiplier tube, exploiting the photoelectric effect. We obtain electron multiplication by accelerating emitted electrons between several dynode stages.

[2] This effect releases electrons with a kinetic energy $W = h\nu - W_0$, when light is absorbed in a suitable material. The Planck constant is h, the frequency of the incoming light is ν and the work of emission is given by W_0. It is important to note that the emission only depends on the **frequency**. The **intensity** only defines the number of generated photo electrons.

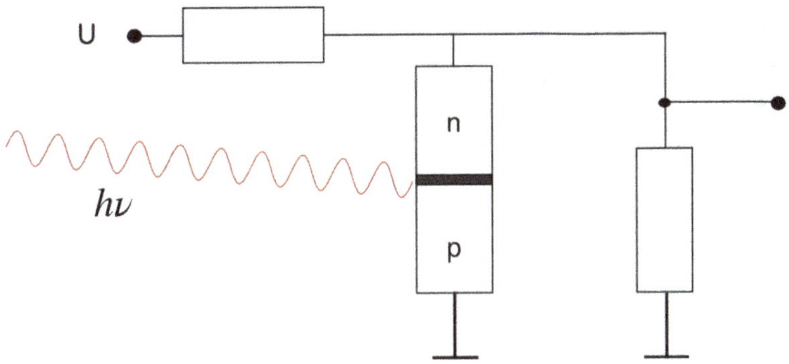

Figure 3.17. Operation principle of an avalanche photo diode in the Geiger mode. The diode is inserted into the circuit in reverse direction and the voltage U is increased above the breakdown threshold for a brief moment. When a photon is absorbed and an electron generated in the p-n junction, a voltage breakthrough is generated.

output electrode. The advantage is the high bandwidth of this detector, but there is a drawback too. The response delay is strongly depending on the number of initial photo electrons. It is important to keep the detection level at the single photon regime to avoid errors from a variable detection delay. More details on this concept can be found in [3]. Since the probe signal propagates at the speed of light, it requires a time tagging device with the resolution of a few ps. It takes 30 ps for the laser pulse to travel a distance of 1 cm. In the early days of the SLR technology, **time interval counters** have been used. The outgoing laser pulse was starting the counter and the incoming signal return stopped the counter and the established time interval corresponded to the time of flight. Satellites in the near-Earth orbit feature roundtrip times of the order of 5–10 ms, The LAGEOS satellites require a signal roundtrip time of ~50 ms and the GNSS orbit already reaches 120 ms. The Moon finally measures roundtrip time intervals of ~2.6 s. In order to use time interval counters on targets that are further away, like the GNSS satellites, the repetition rate of the laser had to be reduced to 5 Hz, which in turn reduced the data yield considerably. Assuming a normal distribution for these range measurements, it is highly desirable to make the data yield high, since the range precision improves with a factor of $\sqrt{n}$, with n the number of valid returns. As a side effect, high-repetition rates make the echo recognition a much faster process and automatic tracking on a low repetition rate is not practical.

Modern timer concepts have switched entirely to the **event timing technique**. Unlike for the time interval counting, this technique can support a high signal rate and therefore may have many shots in flight during the measurement process. Each laser pulse generates a start time tag event from the event timer and each detected return generates the epoch of a stop event. The control program uses the predicted time-of-flight duration in order to associate each stop event with the corresponding start event. Practically all modern SLR systems use event timing and laser repetition rates between several hundred Hz and several kHz.

Laser ranging to satellite targets can be carried out by day and night and even include remote satellites with a very weak return signal, as low as a single created

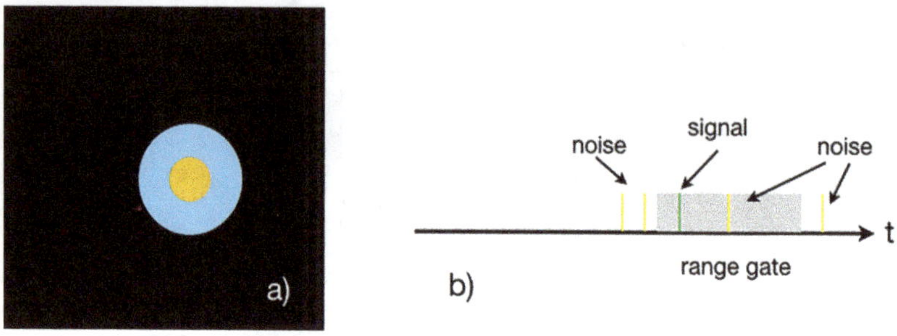

Figure 3.18. In order to obtain even weak return signals from remote targets in bright daylight, the method of spatial (a) and temporal (b) filtering is illustrated in a sketch. The remaining field of view can be as low as 10 seconds of arc and the gatewidth can be as short as 100 ns.

photo-electron in the photodetector. In order to discriminate a valid satellite return against a much brighter level of background light, we make use of three very effective stacked filter methods. Most ranging systems operate a frequency doubled Nd:YAG laser at a wavelength of 532 nm. A narrow optical bandpass filter with a passband of 1–10 nm around the laser wavelength, reduces the background noise already significantly. Even larger than this effect is the reduction of the field of view of the receive telescope by a field stop of the order of 10 seconds of arc. This drastic reduction of the field of view of the telescope sets the requirements for accurate orbit predictions very high. Otherwise, the satellite target will not be visible to the ranging system. The third filter method applies gating of the photodetector that picks up the weak return signal. From the orbit predictions a value of the expected laser roundtrip time is derived. This value and a small acceptance window for the return signal is used to activate the photodetector just prior to the echo signal arrival. Figure 3.18 illustrates the latter two methods in a sketch. It should be noted that the apparent target size in the telescope and the remaining field of view are approximately in the right proportion. This underpins how high the requirements for the orbit predictions and the accuracy of the telescope pointing are actually set for the tracking of such targets in daylight. A tolerance of 100 ns for the gate of a roundtrip signal corresponds to a tolerance for the distance of only 15 m. Although these stringent requirements for the field of view and the detection gate are not applicable to all satellites, these settings are not uncommon.

3.2.4 Orbit predictions, pointing and acquisition

The laser link budget suggests performing laser ranging with a well-collimated beam. The signal level drops off with a factor of R^{-4} with increasing distance (R). A diffraction limited beam at $\lambda = 532$ nm and an aperture of 10 cm has a divergence of just over 1 second of arc. Even when we allow an effective divergence angle of 3–10 seconds of arc, the ranging application requires exquisite quality for the orbit predictions. It also requires excellent pointing and tracking stability of the ranging telescope to keep the satellite in the field of view all the time. Moreover, we actually require a pointing accuracy not far from 1 second of arc absolute. Telescope pointing usually suffers from small instrumental errors, such as encoder offsets relative to true north, nonorthogonality of the horizontal with respect to the vertical axis of the telescope, the leveling error as well as such errors as the nonlinearity of the encoder scale. A **mount model** is usually defined to account for these instrumental short-comings. To set it up, a number of stars with known positions from a star catalogue are observed. Once the pointing offset between the telescope and the true star positions of approximately 20 stars almost evenly distributed across the sky has been observed, a set of linear independent equations is solved and the respective mount model corrections are added as offsets to the commanded pointing positions.

A laser beam with a Gaussian intensity distribution has a beam diameter, which is much larger than any of the optical components in the beam path, such as mirrors or lenses steering the beam. This means that diffraction occurs and the intensity distribution pattern consists of a bright central spot, surrounded by a number of rings of low intensity. This structure is known as the **Airy pattern** and for a perfect flawless imaging system, it defines the spot size or the resolving power of the optical system and corresponds to equation (3.3) above. For the typical laser wavelength of $\lambda = 532$ nm, a 5 cm diameter transmit telescope is already enough to produce a diffraction limited patch of about 10 μrad or 2 seconds of arc, which in turn is small enough to maximize the light intensity at the location of the target, while keeping the target in the field of view, despite the remaining telescope pointing errors.

All this assumes that there is no relative motion between the observing station and the satellite target, but this is not the case. The satellites are moving along their orbit at a considerable velocity. For the satellite LAGEOS, roughly 6000 km above sea level and very roughly 3.5 h for one revolution, we get a velocity of something like 8 km s^{-1}. That means that the satellite has moved a fair bit along its trajectory by the time the laser pulse reaches the target. This effect is known as the *velocity aberration* and can roughly be approximated as [1]

$$\alpha = \frac{2\nu}{c\left(1 - \frac{\nu^2}{c^2}\right)} \approx 2\frac{\nu}{c} \tag{3.4}$$

When we apply this to our example, we obtain an angle offset of 40 μrad or 8 seconds of arc as a worst-case value, since only the tangential component of the trajectory vector on our line of sight matters. So the actual values may differ from one satellite pass to the next.

3.2.5 System calibration

The control system time tags the departure and the return of the laser pulse. Taking the time difference between stop and start pulse provides the roundtrip time of the laser pulse. Most of the travel time is in a vacuum and we recall that the velocity of light $c_0 = 299\ 792\ 458$ m s^{-1} in a vacuum is well defined and the fastest velocity that

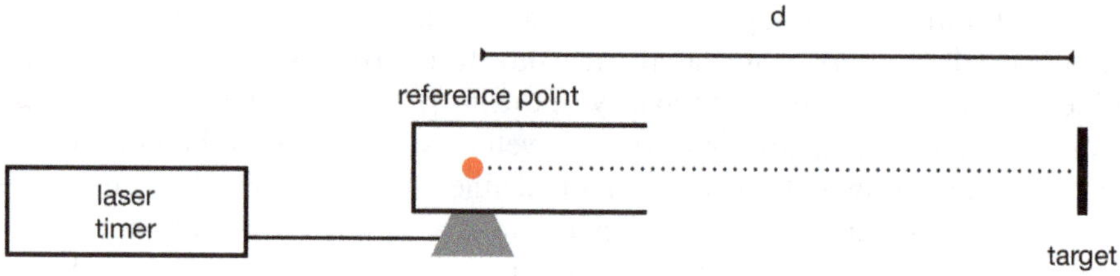

Figure 3.19. Illustration of the calibration process. Since the laser and the timer are not at the system reference point, the extra delay from the internal signal path has to be subtracted out by the help from a local target at a precisely known distance d.

can be realized. Therefore, SLR is a very elegant method to measure distances. This very nice property of time-of-flight measurements, however, comes at a price. Since it takes 30 ps for the laser pulse to travel 1 cm, the required measurement resolution is extremely high.

The actually measured range, however, is longer than the true distance to the satellite. This has two reasons. The velocity of light in a dielectric medium, the atmosphere, is lower than c_0 and we have to compute a correction for this atmospheric delay. Furthermore, there is a not insignificant part of instrumental beam path that the laser beam traverses during the ranging process and which is not part of the geodetic range measurement. In other words, we have to reduce our measurements to the established reference point of the system. This is usually done, by ranging to a local ground target at a precisely known distance. Once the roundtrip time to this calibration target is established with sufficient statistical significance, the internal system delay of the ranging station is obtained by subtracting two times the known distance from the measured roundtrip time interval. Figure 3.19 illustrates this process schematically. Although this process appears to be very plausible, there are a few things to consider in order to avoid systematic measurement offsets. By ranging to a satellite target, the signal levels are small, due to the large attenuation by the distance R, following a R^{-4} intensity dependence. A local ground target does not produce the same attenuation, so the signal levels have to be carefully matched in order to avoid a measurement bias. The wavefront of the ranging pulse at the satellite is flat, but for the calibration at a close distance, there is some curvature. This changes the range, when the pointing position of the telescope varies inadvertently. When more than 1 CCR at the satellite is contributing to the ranging signal, it will distort the echo pulse, which may cause some variable delay in the measured satellite range. This is because the trigger threshold for a laser return usually is constant. So pulse distortions are causing a trigger delay. The high demands on the timing process are in this respect unfavorable. A number of recorded return pulses from the satellite AJISAI, displayed in figure 3.20 show this effect clearly.

This discussion about the proper extraction of the range information outlines how involved this process is, despite its very obvious and simple concept. The stringent timing requirements demand signals, which are as uniform and broadband as possible, with fast rise times. Everything else causes an asymmetry, which introduces systematics and these need to be avoided at all cost.

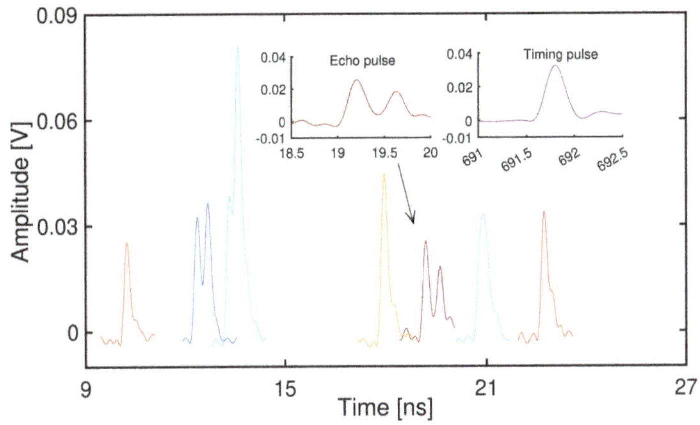

Figure 3.20. A number of digitized different satellite returns from an observation of AJISAI, showing a highly variable signal pattern. A blow-up of one of the echo pulses and a calibration timing pulse are shown in the inset. The local timing pulses in comparison are always uniform. (Figure credit: Jan Kodet.)

3.2.6 Reduction of SLR observations

As stated initially, SLR is a radar technique with extremely precise range resolution. While typical radar operation is associated with a wide swath, the optical satellite ranging has an extremely narrow field of view of only a few seconds of arc. Radar ranges are only approximate and SLR in comparison produces sub-centimeter resolution for distances reaching from the LEO orbit up to the Moon. This narrow field of view requires excellent *a priori* knowledge of the satellite orbit in order to command the required precise pointing of the tracking telescope over the entire path where the satellite is visible. Since SLR can be utilized in daylight as well as in darkness, we require a fairly accurate prediction for each range measurement. The most relevant satellites, which are used for the calculation of the terrestrial reference frame are entirely passive satellites shaped like a cannonball with a number of CCRs on them. They approximate a mathematical dot mass moving in the gravity field of the Earth. A perturbed Kepler orbit is usually numerically adjusted to recent measurements provided by the entire SLR network. This orbit is then extrapolated into the future. With the data density and range resolution of today's network, the predicted orbit is already accurate to about 6 m, which is completely sufficient for the ranging operation. Transformed to the local horizon system of the SLR station, the predictions are providing a list of azimuth and elevation angles and the range as a function of time. The angles are used to point the telescope, while the range information is used to activate the detector just in time before the laser signal returns to the ground station.

When the laser fires at a given epoch, the ranging program adds a time interval of two times the predicted laser travel time together with the refraction correction and the internal laser station delay to the acquired start time. It then subtracts a predefined rangegate open lead time and sets the gating electronics accordingly. Figure 3.21 illustrates this process. A practical example is given by figure 3.22 for the observation of one of the Glonass satellites. The tolerance interval was set to be 1 μs, since at the time of the prediction computation the orbit was not well known.

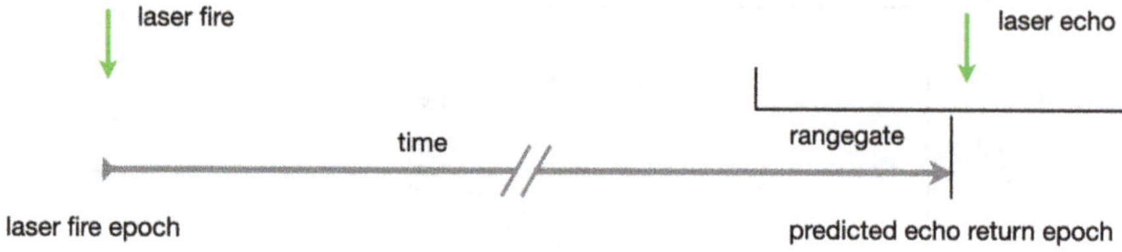

Figure 3.21. Once the laser fire event has been detected, the rangegate generator computes the expected epoch for the satellite return. By activating the detector over a small time window of tolerance around the expected echo epoch, the return signal can be detected even in daylight.

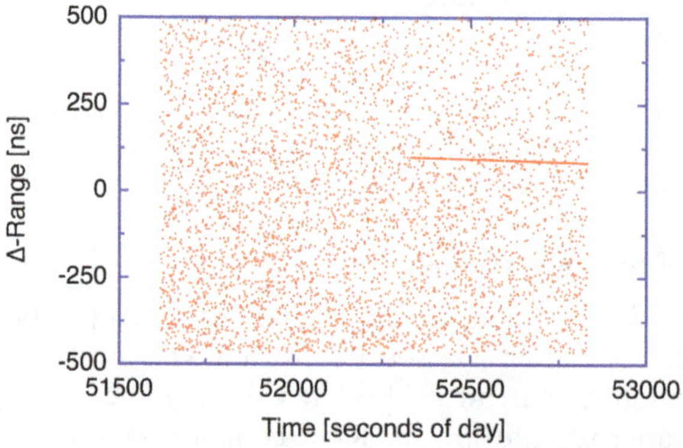

Figure 3.22. A sample daylight observation of Glonass-67. The random dots all over the rangegate are caused by background light. The actual echoes from the satellite reside in the solid track that appears in the second half of the measurement sequence.

After quite some time of stirring the telescope around in steps of ±1–10 seconds of arc around the predicted position in the direction along the track and orthogonal to it, in order to get the satellite into the field of view, echoes from Glonass-67 were finally detected in the second half of the measurement process. While noise detections from the illuminated sky background scattered all over the rangegate appear to be dominating this daylight measurement sequence, the solid line of the actually valid range measurements is much more densely populated. In the postprocessing the noise recordings are removed from the observations. From figure 3.22 one can see that the observed satellite echoes come in slightly later than predicted and they appear with a slope. This apparent drift is caused by an offset between the measurement clock and the predicted time in the orbit calculation. After a short arc orbital fit to the observations is performed, this time and range bias is extracted and the observations are binned into intervals of 5 s for very low earth orbiting satellites up to 300 s for earth orbiters, higher in altitude than 15 000 km. The observations in these bins are then averaged into normal points and reported with the corresponding statistical error. Together with the input of the other techniques of space geodesy, the reference frames can be computed [4]. These normal points are the actual results from laser ranging observation process. Figure 3.23 presents the extracted observations from a pass of the satellite LAGEOS taken at the Geodetic Observatory Wettzell.

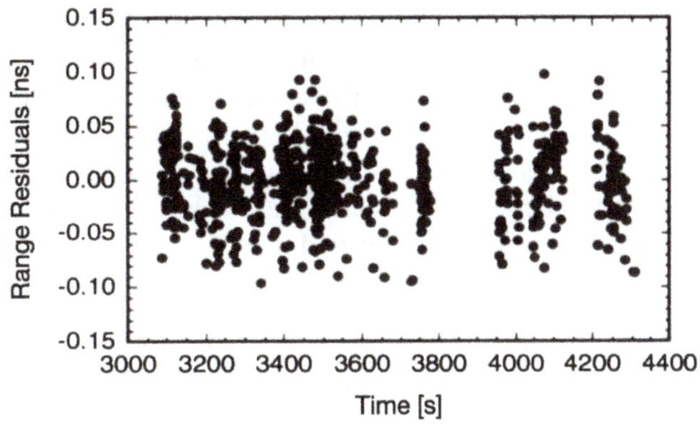

Figure 3.23. Example of an screened observation of a LAGEOS satellite satellite pass. The repetition rate of the laser was 10 Hz. The ranges have been adjusted to the orbit.

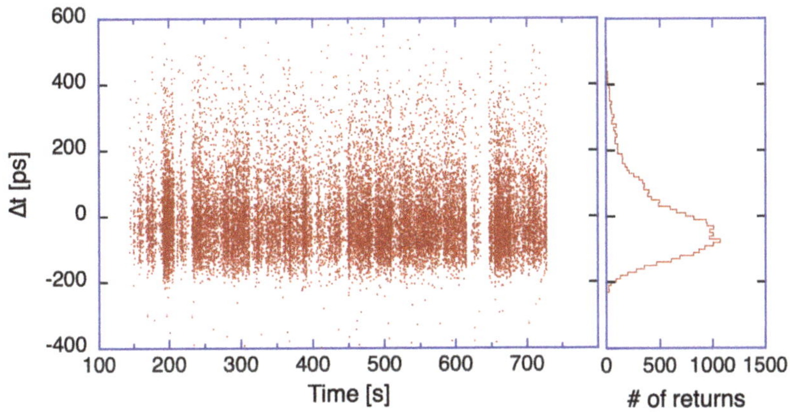

Figure 3.24. Observation of a comparable section of a LAGEOS-2 pass, observed with a fire rate of 1 kHz. The measurements are much denser than in figure 3.23. The histogram on the right illustrates the distribution of the measurements. One can see a slightly asymmetric distribution toward longer ranges.

An obvious feature in this representation is a variable data density. This can have several reasons. One obvious cause for a reduced data rate is additional attenuation, due to thin layers of cloud. In the presence of a thick cloud layer, the data rate drops to zero. This was probably the case in the interval around the second 3900 of the day of the observation. A less obvious, but frequent reason for data loss are errors in the telescope mount model. When these errors exceed 10 seconds of arc, the target is lost from the field of view and it is difficult to acquire it again by commanding offsets to the telescope, due to the low measurement rate.

The laser fire rate plays an important role in the process of SLR return acquisition from a satellite target, since it sets a dwelling time limit for each telescope position in the initial target acquisition process. While most systems were operating at a laser fire rate of around 10 Hz in the nineties of the last century, it reached the level of several kHz in the new millennium. This enabled a much tighter feedback on the tracking loop and proved to be the essential enabler for automatic tracking. The histogram in figure 3.24 outlines another typical issue in SLR very well, namely the nonnormal distribution of the measurements. That means the statistical mean value does not correspond to the true range value. The measurements are slightly biassed, mainly caused by the spherical shape of the satellite allowing several reflectors at slightly different distances to contribute to the return signal. Most cannonball-like satellites cause this effect. A commonly accepted practice is to statistically filter the histogram with a 2.2σ criterion and redo the statistical assessment for a second time.

3.2.7 Daylight operations

SLR detects very low levels of short light pulses, which are reflected off from geodetic satellites with CCRs. In the early days, these observations were restricted to nighttime, in order to avoid saturation of the photodetector. Darkness was also helpful to enhance the signal-to-noise ratio (SNR) of the observations. Better lasers, a significantly improved tracking, accurate orbit predictions and narrow optical bandpass filters allowed the setting of a progressively shorter rangegate, and today there is practically no difference between daylight or nighttime observations. In this context, it is interesting to note that most SLR systems operate in a mode where single photon detection is the norm. Since the orbit predictions are usually excellent today, the station can operate with very short rangegates and sets them such that the rangegate opens about 100 ns ahead of the actual arrival of the satellite echo. In the presence of tight spectral filtering and a small field of view, there are almost no registrations from background noise sources.

3.2.8 Systematic error sources

SLR is the physical realization of a very elegant concept of measuring the time of flight of very short optical pulses with a full width half maximum (FWHM) duration of $10 \leqslant 200$ ps. SLR measurements sample the orbit of small passive satellites, which act like mathematical dot masses. From these orbits the center of mass of the Earth can be established with an uncertainty of less than 1 cm. SLR contributes significantly to the scale of the Earth, leaving an error margin approaching a level of 1 part in 10^9, which is an outstanding achievement. Approximately 30 ranging stations are part of the ILRS, but more than 75% of all measurements come from a group of about 10 productive stations, distributed relatively unevenly across the globe with very few of them in the Southern hemisphere.

The data quality of an individual station can vary significantly over time as hardware ages or new technical components are introduced and it varies also between all the stations. the velocity of light is so large that the optical signal propagates 1 cm in only 30 ps in a vacuum. This puts very high demands on the entire timing process and the utilized instrumentation arrangement. Let us examine this requirement in more detail. Figure 3.25 depicts the ranging process in a *range–time* diagram, with *time* going along the horizontal axis and *range* along the ordinate. The laser fires at t_0. This event is time tagged with a very small uncertainty τ_1, indicated by the bold dot at t_0. The laser pulse propagates toward the satellite and is retro-reflected at the CCRs

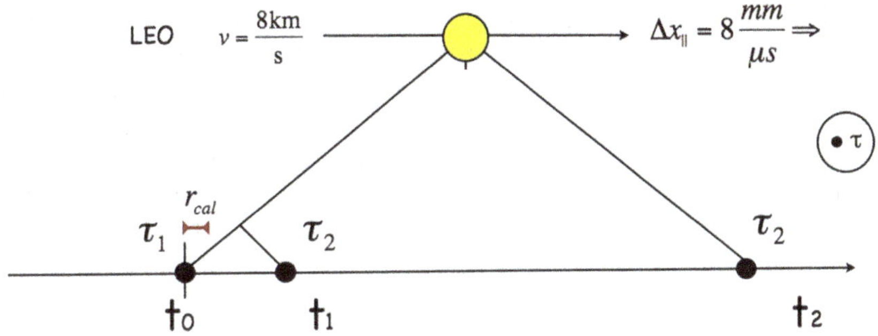

Figure 3.25. Graphical illustration of the ranging process in a range–time diagram and time progressing along the horizontal axis. The yellow circle indicates a satellite in a LEO orbit. The fat black dots mark some undesired unavoidable extra delays, which have to be removed as good as possible.

with no extra delay and returns to the ranging station, where it is detected at some later time t_2, again with a small uncertainty τ_2. Apart from the actual ranging path, which is the distance from the system reference point up to the reflectors at the satellite, the laser pulse progresses through some additional optical path inside the ranging station, both on the way out and on the way back in. We also need to keep in mind that the detection process of the laser signal for the start and the stop signal involves detector delays, cables and timer delays. In order to establish the range to the satellite, the measurement has to be reduced to the system invariant reference point by means of the calibration process as discussed in section 3.2.5 above. With the calibration range r_{cal} accurately known, the respective pair of the start and stop epoch t_0 and t_1 are obtained. It is important to note that the optical and electrical signal propagation path must be as identical as possible inside the ranging station. For the range and the calibration measurement we obtain

$$r(t_0) = \frac{c((t_2 + \tau_2) - (t_0 + \tau_1))}{2} = \frac{c}{2}(t_2 - t_0 + \Delta\tau) \tag{3.5}$$

and

$$r_{\mathrm{cal}}(t_0) = \frac{c((t_1 + \tau_2) - (t_0 + \tau_1))}{2} = \frac{c}{2}(t_1 - t_0 + \Delta\tau) \tag{3.6}$$

Subtracting (3.6) from (3.5) we obtain:

$$r(t_0) = r_{\mathrm{cal}}(t_0) + \frac{c}{2}(t_2 - t_1) \tag{3.7}$$

Some of the laser stations perform *realtime* calibration, while others perform the calibration either before or after the satellite observation. In principle, there could be a shot by shot calibration value for the realtime systems, but it is much more favorable to improve the result by averaging over a larger number of calibration measurements in order to reduce systematic errors by $1/\sqrt{n}$. Note that we have excluded the necessity of the correction from the atmospheric propagation delay from this discussion here for better clarity. This topic will be addressed in section 3.5. Looking again at figure 3.25 it is interesting to note that while we have to perform the timing for the range estimation at a resolution of a few picoseconds, the estimation of the measurement epoch does not have such a stringent requirement. For a satellite moving along its orbit at a velocity of say 8 km s^{-1}, the along track movement corresponds in this example to 8 mm μs^{-1}. Therefore, a timing resolution on the UTC timescale of 100 ns is fully adequate. If, however, highly resolved optical time transfer is required, the start epoch also needs to have a resolution of a few picoseconds. We look at this in more detail in section 5.5.

3.3 Types of satellites and their properties

SLR samples the orbit of a satellite equipped with CCRs. That means it provides precise optical orbit determination, based on two-way measurements utilizing a single clock. Figure 3.26 provides some examples that are discussed in the following. The best results are obtained on small heavy spherical satellites, which have CCRs all over their surface. They resemble a mathematical dot mass and the drag resulting from the resistance of the residual atmosphere is minimal. These satellite targets are carefully designed, so that the center of mass does not change when the sphere is tumbling. The radius between the bottom end of the cube-corners and the center of mass then is a single value correction for the range measurements. For the geodetic community there are a number of such cannonball-type satellites in orbit at around $\approx$ 1000 km orbit height, at $\approx$ 6000 km and at $\approx$ 19 000 km. The satellites are usually smaller when they are in a low orbit, since there is no need for a large ranging cross section. *STARLETTE* and *STELLA* each have a radius of 12 cm and a weight of 48 kg. *LAGEOS* and *LAGEOS-2* are located in an orbit 5000 km further out. They have a radius of 30 cm and a weight of 406 kg. *ETALON-1* and *ETALON-2* are again larger. They have a radius of 64.7 cm and a weight of 1415 kg. This entire group of spherical satellites produces the most accurate orbits and the geodetic reference frames are based on these satellite targets.

Placing a retroreflector somewhere suitable on a satellite body is a fairly simple and straightforward task and this has been exploited by a large number of satellite missions. There are several missions that could be saved by laser ranging because GNSS as their primary tool for orbit determination has failed, in some cases already early in the mission. Usually these satellites do not have a very suitable shape, so the determination of the eccentricity to the center of mass is depending on the instantaneous attitude of the satellite relative to the ground station, which can require a very involved correction model. For these satellites the error margin is larger, but usually good enough for the required purpose.

Another group of important satellites are the large earth observing satellites. They usually operate nadir looking microwave radars for the highly resolved estimation of the current sea level heights. SLR is an important technique to establish an accurate reference orbit, relative to which the sea level changes are obtained. At the same time they are used to calibrate the onboard microwave altimeter. With a current value for sea level rise of 3.8 mm $year^{-1}$, this is a

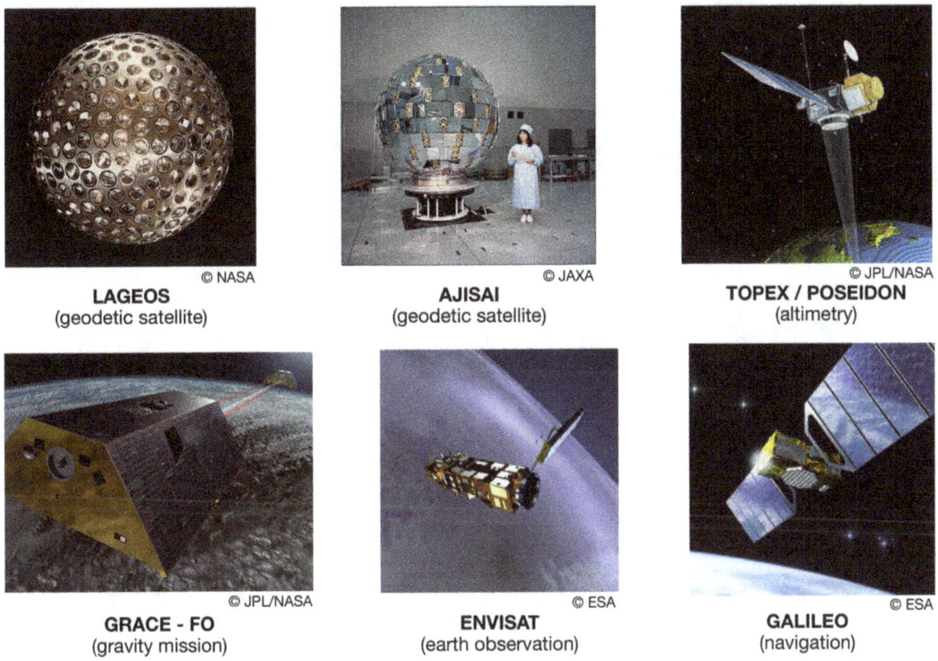

Figure 3.26. Some examples of groups of SLR targets. Apart from the geodetic satellites LAGEOS and AJISAI, there are altimetry missions like TOPEX/POSEIDOM, gravity missions like GRACE-FO, earth observing missions like ENVISAT and the navigation satellites from the GALILEO, BEIDOU and GLONASS constellations.

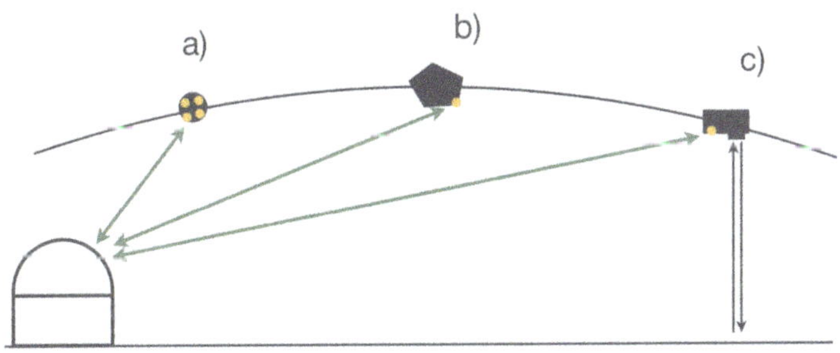

Figure 3.27. Schematical illustration of the different ranging requirements. The cannonball satellites provide the most accurate orbit (a). Bulky satellites that have a reflector cube mounted somewhere, can use SLR as an independent additional method of orbit determination (b). Earth observing satellites use the optical orbits as an accurate reference and for the calibration for the microwave altimeter.

challenging task. Figure 3.27 illustrates these requirements schematically. Until the year 2024, there were several reflector arrays deployed on the lunar surface. They are the targets at the largest distance from the Earth. Two of these targets are flat panels with 100 CCRs, placed on the ground by the astronauts of the Apollo 11 and Apollo 14 mission. A flat panel array of 300 cubes was put on the Moon by

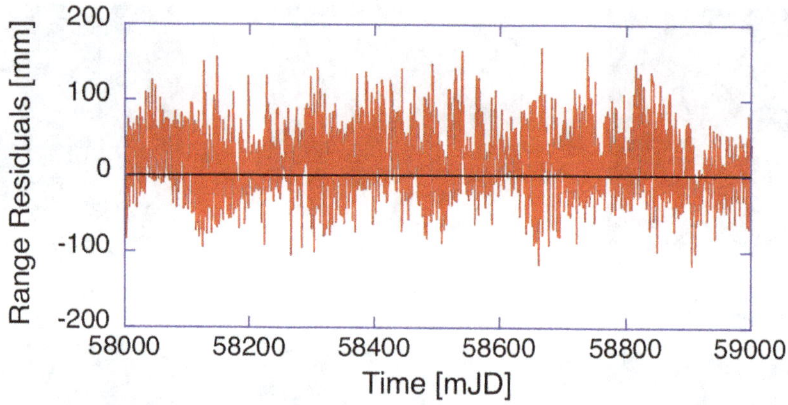

Figure 3.28. Comparison of the microwave derived orbit of Galileo E01 against the SLR derived orbit from the global network. In this example the orbits between these two techniques differ systematically by 2–3 cm. Day 58 000 in the modified Julian date corresponds to the 4 September 2017.

Apollo 15 and two smaller panels are located on the Lunakhod 1 and Lunakhod 2 rovers of the Russian Luna missions.

3.3.1 Microwave orbit versus optical orbit

The comparison of a GNSS derived and an SLR observed orbit usually provides a small systematic disagreement of a few cm. GNSS is a one-way ranging technique, which means that the clock at the satellite and the clock on the ground have to be synchronized by additional observations. Furthermore, small errors in the orbit model, such as the application of an inadequate solar radiation pressure, are perturbing the GNSS orbit calculation. The result is an offset of a couple of centimeters. The example shown in figure 3.28 is taken from the Galileo E01 satellite starting in 2017, which is also equipped with SLR reflectors. There is a consistent bias of about 3 cm apparent in the observations obtained from the global network of observing stations. The GPS constellation only had two satellites supplied with laser reflectors, while the Russian GLONASS and the European GALILEO constellation have cube-corner arrays on all satellites. The ambiguities between clock and orbit, however, are present in all navigation systems. The apparent bias between the microwave and optically derived orbits underpins the importance of SLR in particular for the calibration of altimeter missions and the understanding of small perturbing forces acting on the fairly bulky box-wing GNSS satellite bodies.

3.4 The SLR link budget equation

The link budget of the ranging signal mostly depends on the laser power, the telescope and the properties of the reflector arrays. Atmospheric transmission is of course a factor, but in the presence of a clear sky, the effect is comparatively small. Figure 3.29 identifies the areas of the involved loss processes schematically and they are discussed in the following. A dominant component is the laser itself. It generates

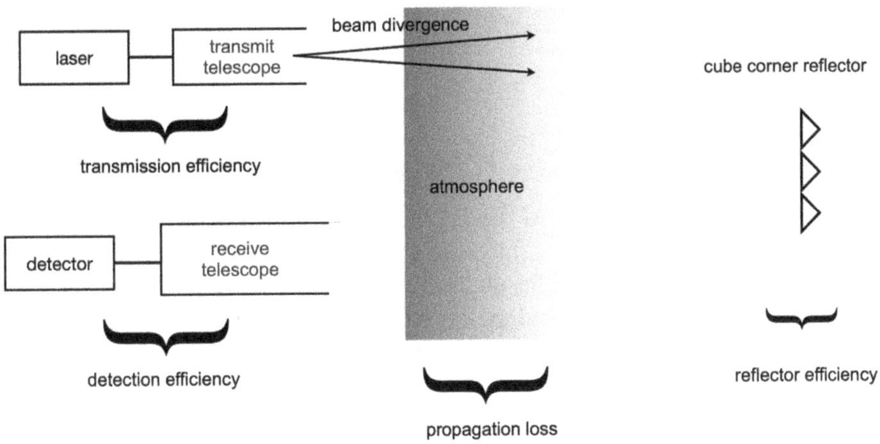

Figure 3.29. Schematic illustration of the loss processes involved in the laser ranging measurements.

the short pulses of energy p_x in accordance with the respective specification. Typical values in the laser ranging community are in the range $1\text{ mJ} \leqslant p_x \leqslant 200\text{ mJ}$. The laser is usually placed in a laboratory separately from the tracking telescopes. Therefore, the laser signal requires a beam guide, usually composed from a number of mirrors, to transmit the probe signal from the telescope. By the time the laser pulse exits the telescope, scattering and absorption losses have occurred at every surface, thus reducing the available laser energy. A total transmit efficiency of 20%–60% is not unusual. The telescope defines the effective divergence of the laser beam. A large divergence of say one minute of arc instead of 2 seconds of arc makes it easier to hit the satellite target, but it reduces the pulse energy at the input aperture of the CCR panel by a factor of 900. This is usually no concern for low earth orbiters (LEO), but prohibitive for high earth orbiting satellites (HEOs). The latter target group requires a narrow beam in order to overcome the losses increasing with the square of the distance, both on the outbound and the inbound path.

The next loss contributor is the reflector. The maximum of reflection is obtained for normal incidence with respect to the front surface with the cross section σ_{cc}, which is given by

$$\sigma_{cc} = \rho A_{cc}\left(\frac{4\pi}{\Omega}\right) = \rho A_{cc}\left(\frac{4\pi A_{cc}}{\lambda^2}\right), \tag{3.8}$$

where ρ is the reflectivity of the reflector, $A_{cc} = \pi r_{cc}^2$ the entrance aperture of the reflector and $4\pi/\Omega$ defines the gain of the retroreflector along the normal vector on the entrance face. Ω corresponds here to the effective spatial angle of the diffraction pattern of the laser beam in the far field. In the essence we are looking here at the gain of a highly collimated beam relative to an isotropic radiator. For a cube corner with a circular entrance aperture the diffraction pattern takes the shape of an Airy pattern

$$\sigma(x) = \sigma_{cc}\left(\frac{2J_1(x)}{x}\right)^2 \tag{3.9}$$

where

$$x = kr_{cc}\sin(\Theta) \tag{3.10}$$

and Θ defines the angle relative to the normal direction on the entrance area of the reflector. The Airy function produces a central patch, surrounded by rings of progressively weaker patches. The angle between the intensity maximum and the first disappearance of the reflected signal intensity is given by the first root of the Bessel function $J_1(x)$. This corresponds to the minimal diffraction limit of an optical system

$$\Theta_0 = 1.22\frac{\lambda}{D_{cc}}, \tag{3.11}$$

where Θ_0 is the half angle of the beam divergence, λ the wavelength and D_{cc} the diameter of the reflector or the telescope aperture. Ω is always very small in all ranging applications. As a consequence σ_{cc} becomes very large. The reflector cross section of a satellite reflector may well reach values of $\sigma > 10^8$ m^2 (figure 3.30). Once reflected, the ranging signal returns through the same path back to be detected by the receive telescope. This adds the divergence loss (R^{-2}) a second time to the link budget. The receiving telescope usually is a lot larger than the transmit telescope in order to collect as much of the returning light as possible. Together with the limited detector efficiency, this reduces the signal intensity even further. The full **link budget equation** in the notation of [5] is then given by

$$n_{pe} = \eta_q\left(E_T\frac{\lambda}{hc}\right)\eta_t G_t \sigma\left(\frac{1}{4\pi R^2}\right)^2 A_r \eta_r T_a^2 T_c^2, \tag{3.12}$$

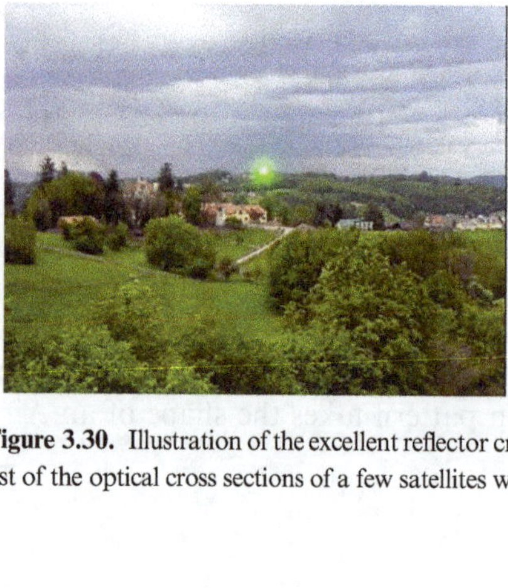

THEORETICAL CROSS SECTION (Million sq m)

SATELLITE	ALTITUDE	CURRENT		REVISED	
			Minimum	Average	Maximum
Starlette	950	0.65	1.00	1.80	2.5
Lageos	6000	7.00	9.00	15.00	23.0
Etalon	19000	60.00	-	55.00	-
Topex	1300	2.00	6.00	33.00	83.0
BeaconC	940	3.60	0.00	13.00	35.0
Ajisai	1400	12.00	-	23.00	-
Gfo-1	800	2.00	.07	.50	1.1
Stella	950	0.65	1.00	1.80	2.5
Jason	1300	0.30	.20	.80	1.7
GPS	20000	40.00	-	19.00	-
Champ	500	1.80	.05	1.00	3.4
Westpac	835	0.03	0.00	.04	.4
ERS	800	0.30	.20	.85	1.6
Glonass396	20000	360.00	-	240.00	-
Glonass132	20000		-	80.00	-
Envisat	800	0.30	.20	.85	1.6
LRE	25000	1.25	-	2.00	-
SUNSAT	600	0.20	.04	.40	1.4

Figure 3.30. Illustration of the excellent reflector cross section of a single CCR at a distance of 6 km (left) and a small list of the optical cross sections of a few satellites with retro reflectors (right). (Photo credit: G Kirchner.)

where n_{pe} is the number of generated photo electrons in the detector and the laser energy E_T is converted to the number of corresponding photons, using the conversion factor (λ/hc) with λ the optical wavelength, h the Planck constant and c the velocity of light. There are several efficiency coefficients in this equation, namely η_q, η_t and η_r. They have dimensionless values between 0 and 1 and describe the normalized efficiency of the detector, the transmission path and the detection path inside the ranging station. The expression in the second bracket is an important component. It allows for the propagation path loss for the uplink and the downlink with R the distance between observing station and satellite target. The reflector cross section σ of the satellite considers the target reflector properties. A_r is the effective area of the primary mirror of the receiving telescope and G_t represents the transmit telescope degree of collimation, called the focus factor. Assuming a Gaussian distribution, G_t can be expressed as

$$G_t(\Theta) = \frac{8}{\Theta_t^2} \exp\left[-2\left(\frac{\Theta_{err}}{\Theta_t}\right)^2\right]. \tag{3.13}$$

As before, Θ is the half angle of the divergence of the telescope and Θ_{err} allows to include the pointing error for the telescope. Although not very evident at first glance, G_t has a profound effect on the calculated link budget, since it can take values well in excess of 10^9, when the collimation approaches the diffraction limit and the pointing error is negligible.

The final two expressions in equation (3.12) are related to the atmospheric transmission T_a and the sub-visible cirrus cloud layer T_c. Most of the laser ranging systems today employ Nd:YAG solid state pulse lasers or something very similar. They operate in the near infrared at the fundamental wavelength of 1.064 nm. Frequency doubling takes them into the visible regime at 532 nm. This has been desirable when the employed detectors exhibited a much higher quantum efficiency in the visible. Since about 2010 this has changed and there are now photodiodes on the market, which exceed a detection efficiency of 70% in the infrared. The atmospheric transmission for 532 nm is about 70% in the zenith direction and drops to about 50% for elevation pointing angles of the telescope of 20°. For the fundamental wavelength the transmission is about 80%, without any dramatic reduction when the telescope is not pointing to zenith. This and the effect that the sky background is colder and therefore less noisy in the infrared regime are making infrared ranging a highly desirable option. Sub-visible cirrus clouds are located in the tropopause composed out of a thin layer of iced up small droplets of water, where most of the resulting crystals are of a size well below 1 μm. At the second harmonic frequency of Nd:YAG they cause more attenuation than for the fundamental frequency of 1.064 nm because the scattering and absorption loss increases as the particle size becomes larger than the wavelength.

In order to provide a feeling for how variable the link budget can be in the day to day operations, we present a comparison of a typical SLR station around the beginning of the millennium under excellent and marginal conditions. The results are presented in table 3.1. The link margin in this example varies over five orders of

Table 3.1. Comparison of the SLR link budget of a typical SLR station around the beginning of the millennium under excellent and marginal conditions [5].

Parameter		Maximum	Minimum
Detector quantum efficiency	η_q	0.18	0.1
Number of photons	$E_T \frac{\lambda}{hc}$	2.7×10^{17} (100 mJ)	1.6×10^{17} (60 mJ)
Transmit path efficiency	η_t	0.66	0.66
Focus factor	G_t	3.2×10^9 (20″)	1.4×10^9 (30″)
Reflector cross section	σ	7×10^6 m^2	7×10^6 m^2
Two-way path attenuation	$(\frac{1}{4\pi R^2})^2$	4.9×10^{-30} m^4	1.2×10^{-30} m^4
		(R = 6000 km)	(R = 8649 km)
Area of receive telescope	A_r	0.4055 m^2	0.4055 m^2
Receive path efficiency	η_r	0.54	0.54
Two-way atmospheric transmission	T_a^2	0.8	0.02
Two-way cirrus transmission	T_c^2	1.0	0.1
Number of photo electrons	p_e	612	0.05

magnitude. Therefore, most of the SLR stations have additional neutral density filters in front of their detector in order to keep the return signal level within one order of magnitude of variation. This is required to avoid systematic measurement errors by exceeding the allowable dynamic range of about one order of signal magnitude. Modern SLR stations differ from this example by utilizing a pulse energy of 1.5 mJ. In compensation they operate on a repetition rate of the order of 1 kHz. Modern detectors offer a quantum efficiency of up to 70% instead of the 18 in the example of table 3.1.

3.5 Refraction correction

The signal path from the ranging station on the ground up to the satellite passes partly through the atmosphere, where the velocity of light is less than c_0. Apart from this additional delay, there is also the effect of refraction bending the beam slightly. The measurement quantity of interest, however, is the straight line distance R, between the reference point of the ranging station and the center of mass of the satellite. This means that the measurements have to be corrected by an adequate model. This requires input data from auxiliary sensors, namely a barometer referenced to the reference point of the ranging system, the temperature and the humidity. We have to note that the barometric pressure along the line of sight is the most significant parameter and it can be established very accurately on the ground. The dependence of the pressure as a function of height above ground is well described by the **barometric height formula** (equation (3.14)), which exhibits an exponential decay as the height increases.

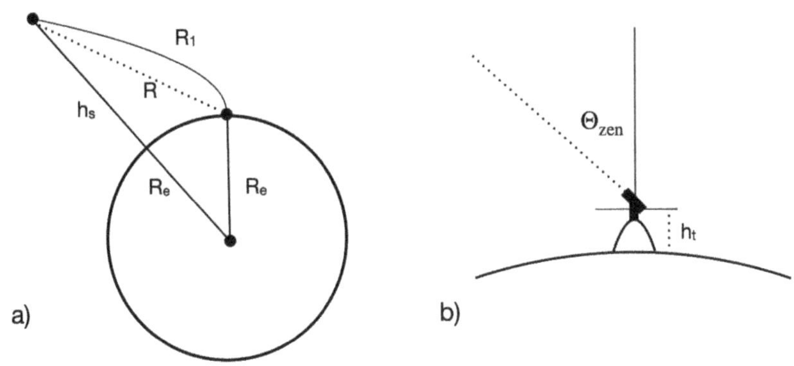

Figure 3.31. Illustration of the effect of the atmosphere on the range measurement. There is a measurement delay caused by the denser atmosphere and a geometrical effect from ray bending. R is the quantity of interest (a) and h_t is the station height above the geoid (b).

$$p_h = p_0 e^{-\frac{\rho_0 g h}{p_0}}. \tag{3.14}$$

The pressure at some height p_h depends on the pressure at the ground p_0, the height h and the gravitational constant g. The temperature is not so well defined and the value measured on the ground is only a rough approximation for the remainder of the atmosphere. However, the refraction correction only weakly depends on it. Similar considerations apply for the humidity, measured at the ground. The refraction correction contains an empirical component for this reason. This has been obtained by adjusting the established corrections with simultaneous *in situ* measurements from radiosonde launches. Figure 3.31 illustrates the problem. With the radius of the Earth given as R_E and the height of the station h_t above the geoid, and the zenith distance Θ_{zen}, the shortest distance, R, between the station on the ground and the satellite is [5]:

$$R = -(R_E + h_t)\cos\Theta_{zen} + \sqrt{(R_E + h_t)^2\cos^2\Theta_{zen} + 2R_E(h_s - h_t) + h_s^2 - h_t^2} \tag{3.15}$$

The atmosphere has the highest density on the ground, which decreases with increasing height. Therefore, it is a reasonable approximation to describe the refraction by a spherical layer model, where each layer has a different refractive index. In practice we find that the delay from a refractive index larger than 1 is the dominant effect. The extra path from the bending of the beam due to refraction is much smaller. This leads to the expression

$$R_l = \int_{r_0}^{r_1} \mathrm{d}r \frac{n_g}{\cos\Theta}. \tag{3.16}$$

The group refractive index is represented by n_g and the angle Θ is derived from **Snellius' law**

$$n r \sin\Theta = n_0 r_0 \sin\Theta_0 \tag{3.17}$$

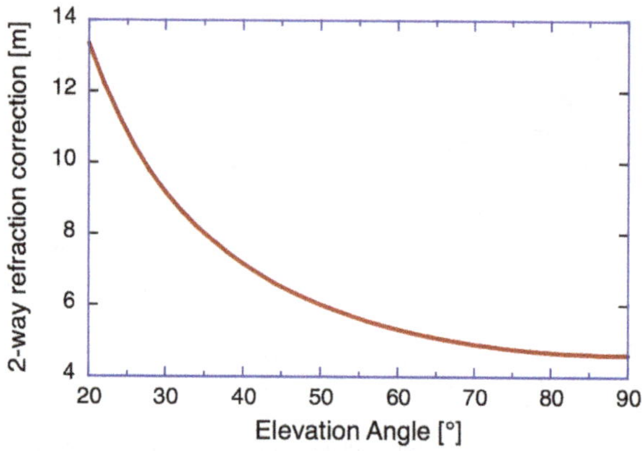

Figure 3.32. The modelled two-way range correction for the passage of the laser pulses through the atmosphere as a function of the elevation pointing angle.

With n_0, the refractive index on the ground ($n_0 \approx 1.000\,272$) we obtain the correction value $\Delta R = R_1 - R$ as

$$\Delta R = R_1 - R = 10^{-6}\int_{r_0}^{r_1} \mathrm{d}r\frac{N_g}{\cos\Theta}\left[\int_{r_0}^{r_1}\frac{\mathrm{d}r}{\cos\Theta} - R\right]. \tag{3.18}$$

Equation (3.18) already contains the commonly used definition

$$n_g = 1 + 10^{-6}N_g \tag{3.19}$$

to account for the very small values involved in the computation. The first term in equation (3.18) contains the effect of the group refractive index, while the second term accounts for the geometrical correction of the bended beam. For a two-way passage of the atmosphere for a ranging laser pulse under typical conditions (temperature = 15.5 °C, humidity = 44% and atmospheric pressure = 952.2 hPa) we obtain a correction value of almost 14 m (see figure 3.32).

To capture the dispersive effect of the atmosphere, we used the expressions [7]:

$$N_g(\lambda) = 80\,343 f(\lambda)\frac{P}{T} - 11,3\frac{e}{T} \tag{3.20}$$

and

$$f(\lambda) = 0.965 + \frac{0.0164}{\lambda^2} + \frac{0.000\,228}{\lambda^4}, \tag{3.21}$$

where P, T and e are the atmospheric pressure, the temperature in Kelvin and the partial pressure of the water vapor. This correction model, known as the Marini–Murray model has been in use for decades to correct SLR data. Many SLR stations today still operate on a wavelength of $\lambda = 0.532\,\mu$m, although there is a progressive trend to move to the fundamental wavelength of Nd:YAG. This led to an effort of updating the correction model, which is now generally in use [8]. Utilizing those for

SLR at the fundamental wavelength of $\lambda = 1.064\,\mu\text{m}$ of the Nd:YAG laser, would make the correction larger by about 20 cm and a zenith distance in pointing of 70°. Evaluating simultaneous measurements of the fundamental wavelength and the second harmonic (dual color ranging) is eventually desired for the correction of refraction induced delays. This has been treated in [12]. At the time of writing none of the SLR stations can provide sufficiently accurate measurements to utilize this correction model.

The atmosphere is not entirely transparent over all frequencies. In the range of ultraviolet radiation below a wavelength of $\lambda = 300$ nm the transmission practically drops to zero. The same applies to the region above $\lambda = 1.1\,\mu\text{m}$. For the second harmonic of Nd:YAG we have a transmission of 80% at a zenith distance of $\Theta_{\text{zen}} = 0°$, which drops to 50% for $\Theta_{\text{zen}} = 70°$. At the fundamental frequency the situation is slightly better by a few percent.

3.5.1 Atmospheric attenuation

The transmission of the atmosphere depends on the absorption of energy in various molecular energy states. If these processes are elastic, it is called **Rayleigh scattering**. In that case the incoming radiation is resonant with vibrational states of the molecules, which in turn reradiate in all directions. In this process the forward scatter is most efficient, followed by the backward scatter. The sideways scatter is again much weaker. The efficiency of this scatter process is proportional to ω^4 and therefore much more effective for higher frequencies A very common observable consequence from this scattering process is the fact that the color of the sky is uniformly blue. The scattered radiation is

$$I_{R_\lambda}(\alpha) = C\sigma_{\text{R}}(\lambda)(1 + \cos^2\alpha), \tag{3.22}$$

where α describes the spatial scattering angle, C is a constant and σ the scattering efficiency. The other prominent scattering process is **Mie scatter**, which corresponds to the scatter of light on tiny particles or aerosols floating in the air. Mie scattering depends only weakly on the radiation frequency with a proportionality of $\approx\omega^{1.3}$, but strongly on the size of the particles. The scatter increases as the particles get larger or denser, which means that the incoming radiation is progressively more attenuated. When the size of the particles is getting close to the wavelength, scattering may increase significantly and reach values of up to two orders of magnitudes more and this does not even include the formation of clouds. For the fundamental frequency $\lambda = 1.064\,\mu\text{m}$ of Nd:YAG lasers, we often find a noticeably lower scatter loss in the atmosphere than for the second harmonic $\lambda = 0.532\,\mu\text{m}$.

The presence of a thick cloud layer absorbs all radiation and renders SLR impossible, which is very obvious and does not need any discussion. Far less obvious is the fact that many places on earth have sub-visible cirrus cloud layers present for 50% of the time, although the sky appears to be clear. Measurements suggest a thickness of this cloud layer of $d \approx 1.3$ km. The transmission of this layer has been established to

$$T_c = e^{-0.14(d \sec \Theta_z)^2}, \tag{3.23}$$

where Θ_z is the zenith distance and the layer thickness d. The laser link equation (3.12) takes this attenuation process into account with the factor T_c.

3.6 The complete observation equation

Our discussion of the laser ranging equation started with the simplified observation equation (see equation (3.2)). A more detailed discussion of the SLR process led to a number of refinements and corrections. Putting them all together provides the complete observation equation, which is written as:

$$\Delta t_e^s = \frac{2}{c}\left(\rho_e^s + \delta\rho_{atm} + \delta\rho_{rel}\right) + \frac{1}{c}\delta\rho_{sys} + \varepsilon_e^s \tag{3.24}$$

The time of flight between the observing station and the retroreflector on the satellite is Δt_e^s. The geometrical straight line distance ρ_e^s contains the extra contribution from the optical path through the atmosphere $\delta\rho_{atm}$ and the relativistic correction $\delta\rho_{rel}$, which is known as the **Shapiro delay**. The system calibration $\delta\rho_{sys}$ reduces the roundtrip measurement to the reference point of the ranging system. Random errors finally are captured in the quantity ε_e^s.

The Shapiro delay [10] accounts for the fact that the propagation of a laser pulse in a gravitational field is delayed. The quantities r^s and r_e amount to the distance of the satellite and the ranging station from the geocenter and ρ_e^s is defined as above. The gravitational constant is G and the mass of the Earth is M. The Shapiro delay is then given as

$$\delta\rho_{rel} = \frac{2GM}{c^2} \ln\left(\frac{r^s + r_e + \rho_e^s}{r^s + r_e - \rho_e^s}\right). \tag{3.25}$$

Typical correction values are in the range of 1–2 cm.

References

[1] David Arnold, private communication

[2] The International Laser Ranging Service https://ilrs.gsfc.nasa.gov

[3] Cova S, Ghioni M, Lacaita A, Samori C and Zappa F 1996 Avalanche photodiodes and quenching circuits for single-photon detection *Appl. Opt.* **35** 1956

[4] Davies P and Blewitt G 2000 Methodology for global geodetic time series estimation: A new tool for geodynamics *J. Geophys. Res.* **105** 11083–100

[5] Degnan J J 1993 *Millimeter Accuracy Satellite Laser Ranging: A Review (AGU Geodynamics Series)* vol 25 (The American Geophysical Union) pp 133–61

[6] Giorgetta F R *et al* 2013 Optical two-way time and frequency transfer over free space *Nat. Photon.* **7** 434–8

[7] Marini J W and Murray C W 1973 Correction of laser range tracking data for atmospheric refraction at elevations above 10 degrees *NASA-TM-X-70555*

[8] Mendes V B, Prates G, Pavlis E C, Pavlis D E and Langley R B 2002 Improved mapping functions for atmospheric refraction correction in SLR *Geophys. Res. Lett.* **29** 53–4

[9] Meschede D 2005 *Optik, Licht und Laser* (Teubner)
[10] Shapiro I I 1964 Fourth test of general relativity *Phys. Rev. Lett.* **13** 789–91
[11] Siegman A E 1986 *Lasers* (University Science Books)
[12] Wijaya D D and Brunner F K 2011 Atmospheric range correction for two-frequency SLR measurements *J. Geod.* **85** 623–35

Chapter 4

Lunar laser ranging

The satellite laser ranging (SLR) technology was discussed in the previous chapter. Here we apply this technology to a test of fundamental physics, namely in gravity. We can consider the Earth–Moon system as an exquisite albeit slightly disturbed dynamic laboratory for gravity. The Moon is going around the Earth in free fall and the Earth itself is going around the Sun in free fall. Nearby planets, our neighbors in the Solar System, as well as minor planets are pulling the orbit of Earth and Moon, changing their relative distance a bit as a function of time. Earth rotation and the rotation of the Moon are complicating the problem further. Furthermore, the Earth and the Moon are not rigid bodies. This means that they deform under the influence of external gravity and this has a significant impact on the measurements too [3]. Figure 4.1 shows a simplified sketch of the measurement arrangement. The observation of interest is the distance of the Earth–Moon system as a function of time between the geocenter and the selenocenter, which is shown in the figure as a red line. However, this quantity is not directly observable. By means of laser ranging we obtain the green line, which is the distance of the reference point of the ground station on earth to the reference point of the lunar reflector on the lunar surface. Furthermore, the range measurement has to be corrected for the effects of refraction. It also has to be corrected for the deformation of the Moon under the influence of gravity, as well as the reciprocal deformation of the Earth by the Moon. To provide an idea, the tidal uplift of the Earth's crust by the gravitational attraction from the Moon is of the order of 30 cm in amplitude, periodically twice a day, ignoring a similar but much smaller effect from the other planets and the Sun for the moment. Polar motion, the Chandler wobble, precession and nutation are complicating things even further. The result is a multiparameter system, partially with external input from other measurements and partially with a nonlinear parameter estimation.

Since lunar laser ranging (LLR) has been in operation since the Apollo era, we have an observation series that is more than 50 years long. The range measurements had a resolution of several meters in the early days, but progressively improved over the

doi:10.1088/978-0-7503-2032-0ch4

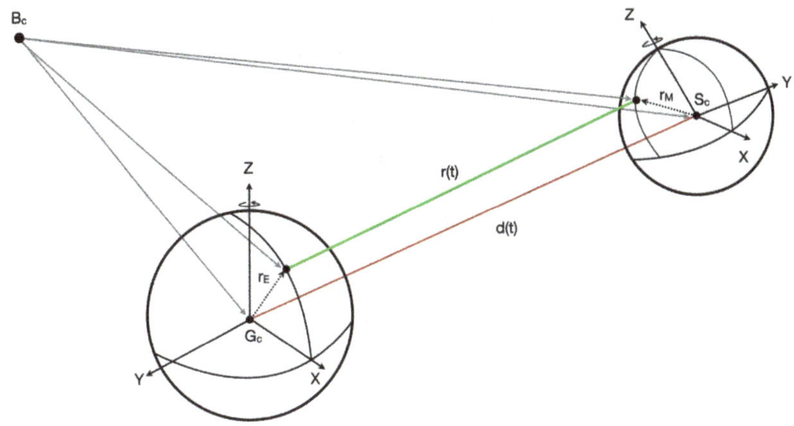

Figure 4.1. Graphical sketch of the basic Earth–Moon distance measurement concept. The measurements are given in a barycentric system. While the distance of interest $d(t)$ is the distance between the geocenter and the selenocenter, the only available observable quantity is the range $r(t)$ as a function of time.

decades to reach a level of a few cm. This makes the numerical adjustment calculation a very involved process [3, 4]. LLR is set to test predictions of general relativity and this technique therefore provides a very high resolution, despite the very many adjustment parameters that go into the calculation. Let us look at an example. Gravitational self-energy is a quantity caused by the gravitational pull of a celestial body on itself under its own mass. Since energy is equivalent to mass, we can divide the mass of the Earth and the Moon into two components, a mass based on self-energy and a mass based on material. If the strong equivalence principle were violated, the lunar orbit would be pulled closer to the Sun by a tiny amount which comes to about 1.2 cm. This effect, if it is indeed present, is known as the **Nordvedt effect** [6]. The current limit after more than half a century of observations is modeled as

$$\Delta r_{\text{earth-moon}} = 13.1\eta \cos(\omega t) \text{ [m]} = A\cos(\omega t), \tag{4.1}$$

where ω is the synodic frequency or expressed differently the angular velocity of the Moon around the Earth. Relativistic tests deal with exceedingly small numbers as we can see from this example. The current limit on the Nordvedt effect is $|\eta| \leqslant 3 \times 10^{-4}$ [3]. Another theory to test would be the weak equivalence principle, a violation of which would also cause a small differential shift in free fall of Earth and Moon toward the Sun. The Earth dominantly contains a large iron core, while the Moon is mostly composed out of silica. If there is something like a material dependence of gravity, it would become visible in the Earth–Moon distance. The amount would be similar in size, compared to the Nordtvedt effect. Over the years there have been a number of theories that postulate the possibility of a secular change in the value of the gravitational constant G. If that is so, it would have an effect on the lunar mean distance, visible in a value of $\dot{G}/G$. This is another prediction that can be tested with LLR. Since the tidal friction has a similar signature, one can observe the time evolution of the solar longitude to constrain the probably existing rate of change. The long time series of LLR observations would show a potential tiny rate of change. At this point in time, the observations set a stringent limit of three parts in 10^{13} for $\dot{G}/G$.

From the brief introduction above, we have seen that our Solar System can be considered an immense sensitive gravity laboratory and the list of testable theories is not even complete. Tests of general relativity, such as gravitomagnetism, geodetic precession and post-Newtonian parameterization can make use of LLR observations, which already show a fantastic resolution. At the time of writing, no violations of the established theories have been found yet. The main effort today is directed to making the time-series of observations longer, improve the resolution and in particular to reduce systematic errors. In the next section we take a closer look at the technology of LLR.

4.1 The LLR observation technology

From a conceptual point of view, SLR and LLR are fairly similar. Both techniques utilize time-of-flight measurements for the orbit determination of artificial and natural bodies in space, using short pulse lasers and retroreflectors. While the SLR satellites are built for this purpose, the laser reflectors for LLR have to be placed onto the Moon. At the time of writing, there are five cube-corner reflector arrays and one single cube reflector on the Moon. Three of these targets have been placed on the ground by astronauts and the three others have been deployed robotically. Two of the latter arrays are mounted to the Lunakhod rovers and the single reflector cube was placed onto the Blue Ghost lander, which came down in the Mare Crisium toward the right limb of the Moon on 2 March 2025. When we look at figure 3.3, we see examples of typical geodetic cannonball like satellites. These targets have the advantage that the way they are orientated in space is irrelevant, while they orbit around the Earth. Only the reflectors that are orientated toward the ranging station are contributing to the time-of-flight measurement. The satellite 'STARLETTE' on the right is in a lower orbit, hence it is smaller and fewer cube-corner reflectors contribute to the measurement. 'LAGEOS' on the left is in an approximately six times higher orbit. Therefore, it is larger and several reflectors, namely the first ring around the central cube looking straight at the ranging station contribute to the return signal. We recall that the return signal intensity drops off with the fourth power of the distance between measurement station and target. Furthermore, for these satellites it is simple to connect the range measurement to the center of mass of the satellite.

On the Moon things are very different. The cube-corner reflectors only make up a very small spot somewhere on the lunar surface. In addition to that, all but one reflector are composed out of several, if not many, cubes. The Lunakhod rovers house arrays of 14 triangular cube-corner reflectors, Apollo 11 and Apollo 14 have placed flat panel arrays on the ground, which contain 100 reflectors, while Apollo 15 positioned an even larger array with 300 reflectors on the ground at the landing site (figure 4.2).

Figure 4.2. 21 July 1969: Edwin (Buzz) Aldrin carries the seismometer (left) and the laser reflector to the deployment location (left). The fully installed laser ranging panel is shown on the right (figure credit: NASA).

The first problem for the measurements are the reflector positions. The panels were just placed to the lunar surface. By the time that the Apollo 11 team headed home from the Moon, it was not known, if the reflector would ever be found from the ground station laser beam and if lunar returns would ever be recorded. The Apollo lunar ranging project had prepared two ground stations for the ranging measurement, namely the Lick Observatory in California and the McDonald Observatory in Texas. One wanted to make sure that at least one ground station was available for ranging, due to uncertain weather conditions. Today it appears a little like an irony, but there were only 4 days allocated for the initial ranging attempt. Observing time on large telescopes was in high demand at that time and the astronomers apparently did not have much faith in that experiment. On the last of these four days and eleven days after the deployment of the reflectors, the Lick Observatory provided the first range measurements, approximately 16 single echoes, McDonald was not operated at that time, due to thick cloud layers. Although the proof of the existence of a working LLR array was provided by Lick, all the early science was done at McDonald. In the beginning on the large 2.7 m main telescope, later on a much smaller but dedicated system nearby. While McDonald essentially was the only station to provide LLR measurements throughout the seventies, another dedicated station was set up for this purpose in Grasse (France, Coté d'Azur) and started operations in 1981. Since then this station turned out to be the most productive LLR site until today. Over the eighties and nineties there were some minor contributions to LLR from stations on the Haleakala (Maui, Hawaii), Orroral (Australia) and Wettzell (Germany). After the change of the millennium, two more stations started operations. One was located at Matera (Italy) and the other on the Apache Point Observatory (USA, New Mexico). While the former only provided occasional observations, the latter station became an important partner in the network, because of the high quality of their observations. To understand this better we need to look deeper into the technical side of LLR.

For convenience we reproduce the link budget equation (3.12) from the previous chapter, which of course applies to LLR in the same way.

$$n_{pe} = \eta_q\left(E_T\frac{\lambda}{hc}\right)\eta_t G_t\sigma\left(\frac{1}{4\pi R^2}\right)^2 A_r\eta_r T_a^2 T_c^2, \tag{4.2}$$

However, in order to perform LLR we have to take a closer look at a few parameters. While SLR is fairly forgiving, LLR is not. First and most importantly we need to consider the R^{-4} distance dependence of the link budget. While satellites are usually in close earth orbits, namely between 600 and 20 000 km, the lunar targets are between 357 000 and 407 000 km away. Since the Moon is in an elliptical orbit the distance is variable over a lunar month. But scaling the power drop off over this distance by the power of 4, causes a considerable power loss of 18 orders of magnitude. There are only two factors in equation (4.2) that can improve the link budget again. These are the quantities G_t and σ. The former is the transmitter gain and the latter is the reflector cross-section. According to [1] this transmitter gain is given by

$$G_t(\theta) = \frac{8}{\theta_t^2}e^{-2\left(\frac{\theta}{\theta_t}\right)^2}. \tag{4.3}$$

In this equation θ describes the far field divergence half angle, between the center of a Gaussian beam and the $1/e^2$ intensity drop off (corresponding to 13.5% of the maximum value), while the angle θ_t accounts for the pointing error to the target. This term describes the gain of a tightly focussed beam over an isotropic radiator. LLR requires a beam divergence not exceeding 2 seconds of arc. At the same time it means that the telescope pointing error must not be larger than 2 seconds of arc either. We can conclude that the location of the target must be precisely hit and that the collimation of the beam must be as tight as possible. For the example above G_t yields a value of 1.15×10^{10}, making up for some of the distance dependent signal loss. A beam with these transmit parameters still generates a patch of light with a radius of 3.7 km on the Moon. The pointing error deserves another discussion. It is not only acting on the telescope steering, but also includes the detrimental effect of atmospheric seeing. While the sites at Apache Point and McDonald usually feature excellent seeing conditions with values below 1 second of arc for the turbulence, the Wettzell site experiences values more like 5 seconds of arc, which diminishes G_t already by 30%.

Another important effect in the link budget equation is site independent and that is the optical cross-section of the reflector. In this discussion we only want rough order of magnitudes to illustrate the effect and we take equation (4.4) from [1].

$$\sigma = \rho A_{cc}\left(\frac{4\pi}{\Omega}\right) = \rho A_{cc}\left(\frac{4\pi A_{cc}}{\lambda^2}\right) \tag{4.4}$$

For the lunar reflectors we get the numbers as shown in table 4.1 for σ as a rough order of magnitude estimate. These values apply for the situation when the laser

Table 4.1. An idealized comparison of the optical cross section of all the lunar targets, based on the theoretical cube-corner performance as per equation (4.4).

Lunar target	Cross-section (m^2)
Apollo 11	1.4×10^9
Apollo 14	1.4×10^9
Lunakhod 1	4.1×10^9
Lunakhod 2	4.1×10^9
Apollo 15	4.0×10^9
NGLR-1	4.0×10^8

beam hits the cube-corner exactly square on. It does not account for any degradation of the reflectors due to thermal bending caused by solar radiation, which can have a profound effect on the efficiency of the reflector. The efficiency of the Lunakhod 2 panel is usually between the Apollo 11 and Apollo 15 reflector, when the Lunakhod rover is in darkness. When the rover is illuminated by the Sun, the efficiency drops to values below the Apollo 11 target. This has been quantitatively examined in [2]. Another effect of degradation is the potential partial coverage of the lunar reflectors by electrostatically moving lunar dust, which can be responsible for the observed factor of 10 lower efficiency of the lunar detectors [5]. From this discussion we can see that efficient cube-corner reflectors and a well-focused beam can make up for the extremely high losses in lunar laser ranging. Nevertheless, it is a very challenging project and the lunar distance is about the maximum distance that can be covered by two-way ranging in a meaningful way.

Now let us look at the newly deployed reflector NGLR-1. Surprisingly, it is one order of magnitude lower in efficiency than the earlier ones. It does not seem desirable, since a novel reflector should, if anything, make it easier to track a lunar target. This leads us to a property of lunar targets that is not present for the cannonball type of satellites in a closer earth orbit. The Moon is in locked rotation around the Earth and this is the cause why we always see the same side of it. Although trapped in a gravitational well, the Moon still appears to bounce around a little bit, so that this tumbling motion turns the Moon by nearly 8 degrees. This effect is caused by the elliptical orbit of the Moon and also on the observer's location, which is not on the direct line of sight between the geocenter and the selenocenter. A nearly eight-degree relative rotation between a LLR ranging station and the lunar reflectors has a profound effect on the reflector efficiency.

We recall that the Apollo reflectors use flat panels with 100 or 300 cube-corner reflectors. When the angle of incidence coincides with the normal direction, the efficiency of the reflector is higher and the temporal spread of the reflected pulse is short. However, when the Moon tilts sideways by a few degrees due to libration, the efficiency of the reflector reduces and the spread of the return pulse increases. This

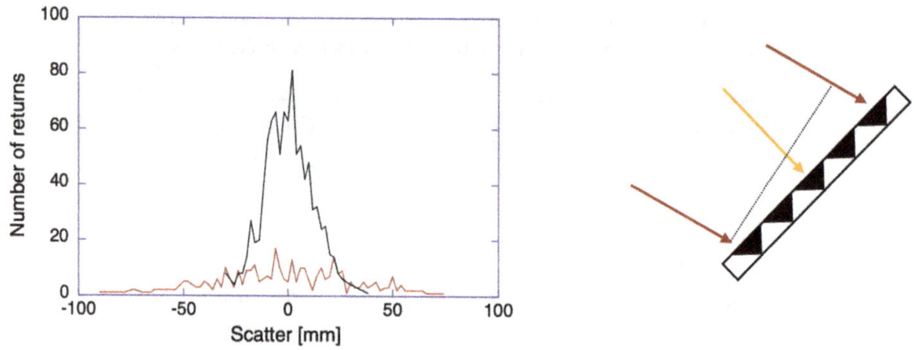

Figure 4.3. A histogram of LLR measurements taken at the Geodetic Observatory Wettzell under low (black curve) and high (red curve) lunar libration. Under high libration the return rate reduces significantly, while the spread of the measurements increase. On the right side there is a sketch of a lunar panel reflector. Under normal incidence (central arrow) all the individual reflections add up coherently. With increasing libration, the reflections progressively develop a path delay, which causes the signals to spread and the observed reduction in signal.

effect is illustrated graphically in figure 4.3. The diagram on the left side shows measurements of the Wettzell station, which were taken under conditions of low and high libration. As the angle of incidence is moving away from the normal direction, the efficiency of the reflectors decreases significantly. At the same time the scatter of the measurements increase. The cause for this larger scatter is illustrated on the right side of the figure. Due to the flat panel design of the reflector, there is a geometrically induced delay between different rows of the reflector. The individual return pulses do not stack up on top of each other, as they do under normal incidence. Instead, there is a little extra path length from one row to the next, which broadens the return pulse. To make the ranging measurements independent of the detrimental effects of libration, it was necessary to go away from the flat panel design and to develop a sufficiently efficient single reflector for LLR applications. From equation (4.4) we can take that the efficiency of the reflector increases with the power of 4 as the radius of the reflector area goes up. Together with considerations for thermal deformation and payload weight, this led to the design of the NGLR-1 single cube-corner reflector.

Figure 4.4 depicts a prelaunch photo of the 10 cm diameter NGLR-1 reflector. When it was put onto the Moon with the Blue Ghost lander on 2 March 2025, it demonstrated the expected improvement for the LLR application. The comparison between a measurement session to Lunakhod 2 and NGLR-1 taken in the same night by the WLRS confirms the expected improvement and it also shows the expected lower efficiency of the new reflector. There are considerably more lunar echoes recorded from the Lunakhod 2 reflector and they scatter three times as much as the measurements from NGLR-1.

Now that we have looked at the reflector efficiency and orientation properties more closely, we can understand why it either required a sufficiently large and

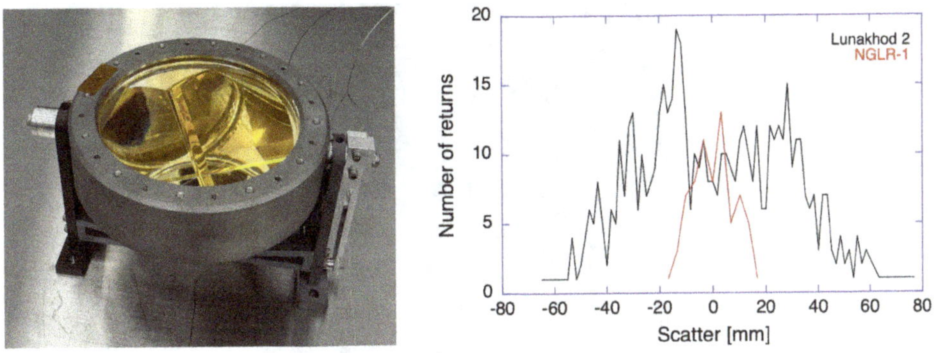

Figure 4.4. A prelaunch photo of the NGLR- reflector, which features an aperture diameter of 10 cm (left). It was put onto the Moon on 2 March 2025. A comparison of the scatter between the Lunakhod 2 and NGLR-1 reflectors from the same night is shown on the right side. Although less sensitive, the NGLR-1 provides much sharper range measurements. (Photo credit: D. Currie.)

rugged rover or a manned mission to put the first lunar reflectors on the Moon. The tolerance for the positioning of the arrays is only a few degrees, preferably even less than that. This was not within the technological range of the landers in the late sixties. Furthermore, we may wonder why NASA did not use more laser power in order to take reflections straight from the lunar surface for ranging. When we compare the optical cross-section from a 2″ divergent laser beam on the lunar surface against the Apollo 11 reflector as an example, assuming an average albedo of 15% we get a value of 6.5×10^6 m^2 for the ground echo, which is about a factor of 200 times less than the efficiency of the Apollo 11 panel. The illuminated patch on the Moon would cover an area of almost 14 km^2 and this raises two concerns. First of all, what is the exact location of the reflection point. We need to have well-defined and stable coordinates for the reflector on the Moon, as you can see from the illustration in figure 4.1. The other concern is about the terrain roughness. The Apollo 11 panel features an edge length of 46 cm. A libration angle of 7° already produces a reflection delay between the first and the last row of almost 6 cm. In order to be compatible, the illuminated lunar patch would have to be flat with a remaining roughness not larger than 6 cm. Such a condition does not exist. Including beam steering effects from atmospheric seeing, which causes the laser beam to bounce around on the Moon, the returning pulses would spread out in time by several meters, provided they could be detected. In fact, this kind of measurement had been tried before the Apollo landing and ground echoes had been detected. In terms of scientific applications, these measurements did not produce useful results. So this effort was abandoned quickly. In the early nineties the Wettzell Laser Ranging System joined the LLR operations. The final figure in this section shows the system on the night of the first returns in October 1991 (figure 4.5).

Figure 4.5. The Wettzell Laser Ranging System on the night of the first successful lunar observation.

References

[1] Degnan J J 1993 *Millimeter Accuracy Satellite Laser ranging: A Review (AGU Geodynamics Series* vol 25 (The American Geophysical Union) pp 133–61

[2] Goodrow S D and Murphy T 2012 Effects of thermal gradients on total internal reflection corner cubes *Appl. Opt.* **51** 8793–7

[3] Müller J, Murphy T, Schreiber K U, Shelus P J, Torre J M, Williams J G, Boggs D II, Bouquillon S, Bourgoin A and Hofmann F 2019 Lunar laser ranging: a tool for general relativity, lunar geophysics and Earth science *J. Geod.* **93** 2195–210

[4] Murphy T W 2013 Lunar laser ranging: the millimeter challenge *Rep. Prog. Phys.* **76** 076901–21

[5] Murphy T W, Adelberger E G, Battat J B R, Hoyle C D, McMillan R J, Michelsen E L, Samad R L, Stubbs C W and Swanson H E 2010 Long-term degradation of optical devices on the Moon *Icarus* **208** 31–5

[6] Nordtvedt K 1982 The fourth test of general relativity *Rep. Prog. Phys.* **45** 631–51

IOP Publishing

The Measurement Techniques of Space Geodesy
Optical technologies
Karl Ulrich Schreiber

Chapter 5

Laser ranging using transponders

In the previous chapter we saw that the lunar distance is about the maximum distance that can be covered by two-way laser ranging. This is due to the immense divergence loss, which increases by R^{-4}, with R the distance between transmitter and reflector. So, if we wish to extend the range of accurate time-of-flight measurements, we have to utilize one-way signal propagation. Apart from reducing the propagation loss to R^{-2} this has an immediate consequence, namely the systematic effects of the clock will no longer cancel out. We recall that the transmission of ultra-short laser pulses has a very wide bandwidth. In other words, the optical transponder technology is very suitable for high data rates and high-resolution time transfer. In this chapter we look at several applications for transponder ranging.

5.1 The concept of transponder ranging

Since the optical wavelength is much shorter than microwaves, the diffraction limit ($\theta = \lambda/D$) is also much lower and this is an advantage in favor of optical deep space communications. A transmit/receive package, containing an optical telescope, a laser, a modulator and control electronics can be compact and lightweight, making them suitable for satellite operations. Optical communication systems can work on the detection of single photons. That is a significantly lower power limit than what microwave receivers can work with. With all these advantages, there is also the disadvantage that much higher demands on the telescope pointing have to be satisfied. This problem can be mitigated in practical applications by providing tracking support with laser beacons. We can distinguish several types of transponder operations [3].

- **Laser beacon**: Modulated or pulsed laser sources, whose code signature is derived from the satellite clock.
- **Pulse-answer systems**: A photodetector detects a signal from an interrogation laser and fires a signal back.

doi:10.1088/978-0-7503-2032-0ch5

- **Asynchronous one-way ranging systems**: There are two one-way ranging systems looking at each other and firing without any synchronization. The distance is computed retrospectively by comparing the recorded fire and detection epochs of both systems.
- **Optical time transfer**: The recorded signal detections on both ends allow a clock comparison.
- **Optical communications**: Apart from pulses there can be code transmissions or a mixture of both, so that both timing pulses and data signals can be used for comparison, but also information transfer.

Figure 5.1 illustrates the simplest form of a transponder concept, namely the principle of a pulse-answer system. A transmitter sends out pulses that attenuate with R^{-2} until they hit the transponder. When the pulses are detected, the transponder in turn fires pulses back toward the transmitter. We note that the process of detection and transmission of the return signal is not instantaneous. Therefore, there is a time interval τ, which we either have to make technically constant or measure along with the transmission and arrival times. Ignoring the atmosphere and the corrections for the reference points in this discussion, we obtain the range between the two systems as

$$r = \frac{c(t_{A2} - t_{A1} - \tau)}{2}, \tag{5.1}$$

where t_{A1} is the laser fire epoch of the ranging system 1 and t_{A2} the detection epoch of the return pulse from system 2. In this case, τ is assumed to have a constant value. If it would be variable, it would be required to measure it and to transmit the respective delay by telemetry. This would complicate things considerably. However, it would potentially have a lower error margin, since it is a great challenge to keep τ absolutely constant. The maximum measurable distance depends on the size of the telescopes, the divergence of the respective laser beams and the laser power itself. While lunar laser

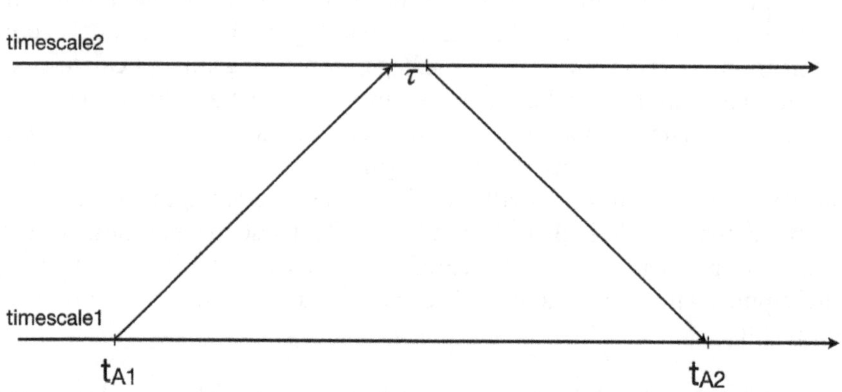

Figure 5.1. Illustration of a simple pulse-echo transponder function in a time–range diagram. Time runs horizontally and the distance is indicated in the vertical direction. The little gap indicated with τ in time scale 2 marks the required delay for the system to respond.

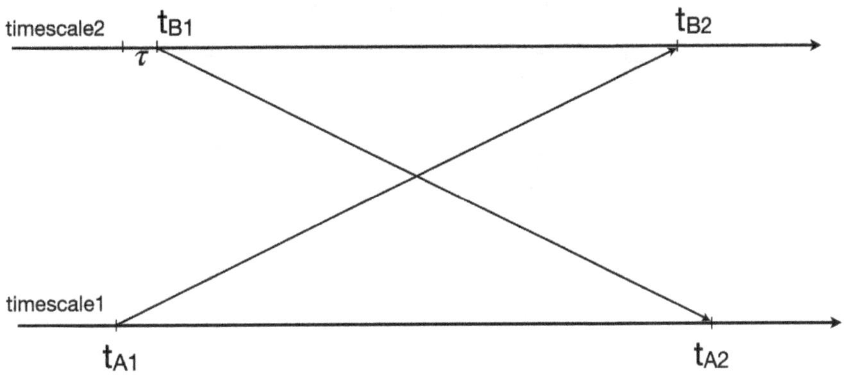

Figure 5.2. Illustration of the asynchronous transponder arrangement. Station 1 records all transmit (t_{A1}) detection (t_{A2}) events, while the transponder records the respective t_{B1} and t_{B2} events and sends them per telemetry to Station 1. By combining the two sets, the distance and the offset τ in the time scales is obtained.

ranging covers nearly 400 000 km distance in two-way, this kind of transponder extends the distance limit to several million km as we will see in section 5.2.

When two transponder ranging systems look at each other, it is not necessary that they operate synchronously, which would limit the repetition rate to the roundtrip time. A much higher data rate is achieved by asynchronous operation. In this case a telemetry link between the two ends of the ranging transponders is required. Both ends fire their laser and record the transmit-epoch sequence. Furthermore, they time the detection epoch of all signals that enter the receiving telescope. After merging the two datasets, the distance and clock offset can be reconstructed as functions of time as well as the offset between the two involved time scales. Figure 5.2 illustrates the approach. From the combination of the measurements [3] of Station 1 and Station 2, the distance is obtained as

$$R = \frac{c}{2}(\delta t_{BA} + \delta t_{AB}) = \frac{c}{2}[(t_{A2} - t_{A1}) + (t_{B2} - t_{B1})] \tag{5.2}$$

and the offset between the two time scales is

$$\tau = \frac{[(t_{A2} - t_{A1}) - (t_{B2} - t_{B1})]}{2\left(1 + \dfrac{\dot{R}}{c}\right)}. \tag{5.3}$$

The asynchronous transponder application requires an additional communication channel in order to collect all measurements from both ends for the evaluation. However, the advantages outweigh this extra effort. The repetition rate can be much higher than the roundtrip time in the synchronous transponder concept.

5.2 Examples of transponder missions

Early efforts to evaluate the potential of transponder applications were carried out as early as 1992, when laser pulses transmitted from Earth were identified on camera

Figure 5.3. Example image from the GOPEX experiment, where the GALILEO spacecraft detected laser pulses transmitted from Earth at a distance of 1.4 million km. (Photo credit: NASA/JPL.)

images of the GALILEO space probe en route to Jupiter at a distance of 6 million kilometers [1]. The camera was scanning slowly during the experiment. The recorded laser pulses from the Starfire Optical Range of the U.S. Air Force Phillips Laboratory near Albuquerque in New Mexico at 10 Hz rate are shown in figure 5.3. The second trace to the left was generated by laser pulses from the Table Mountain Observatory of the Jet Propulsion Laboratory near Wrightwood in California at a rate of 15 Hz.

The laser stations were operated on the night side of the Earth. The terminator region is visible on the right side of the image. The GOPEX experiment (Galileo Optical Experiment) tested the uplink pointing requirements for future optical communication data links. Ranging was not possible at that time. The second experiment of opportunity came in September 2005 when the Mars Global Surveyor (MGS), orbiting the planet Mars in an orbital height of 400 km, became available for a three-day experiment. The laser ranging subsystem 'Mars Orbiter Laser Altimeter' (MOLA) had mapped out the Mars topography over several years and the laser was no longer working. MOLA featured a 50 cm aperture telescope and the ranging detector with a field of view of 0.85 mrad was still operational and used as a photometer. For the interplanetary one-way ranging experiment, the MGS spacecraft was turned, so that the MOLA telescope was roughly facing Earth. This happened while Earth and Mars were approaching conjunction with a distance of about 80.1 million kilometers. The ground station at the Goddard Geophysical and Astronomical Observatory (GGAO) in Greenbelt had a 1.2 m aperture transmit telescope and a pulse energy of 90 mJ. The laser repetition rate was 49 Hz and sequences of 6 pulses 'on' and 6 pulses 'off' corresponding to a pulse spacing of 20 ms were transmitted. This pattern was chosen to make the identification of the laser detections unambiguous. The detection window of the MOLA receiver is 125 ms long, so that up to six events within that window can be counted.

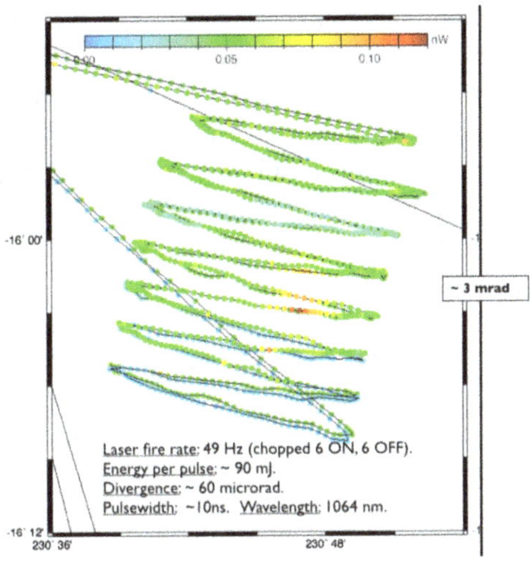

Figure 5.4. An example scan of the MOLA detector acting as a photometer, while the MGS spacecraft is slowly scanning through an area in the direction toward Earth (courtesy: J McGarry, GSFC).

The MGS spacecraft was set into a slow oscillating motion, scanning the Earth direction within a window of 3 mrad. Using the MOLA detector as a photometer, the direction toward Earth could be obtained. This is illustrated in figure 5.4, where blue to green colors indicate a low noise count rate and yellow to red colors indicate a higher noise count rate. From the mapped out region, there were three scans that passed over the Earth during the scan pattern. The ground station used a number of geodetic satellites to verify the correct on axis pointing of the transmit telescope and a laser divergence width of $\approx$100 μrad was established. On 28 September 2005, laser pulses were counted successfully when the field of view of the MOLA receiver swept across the Earth when performing the scan pattern, demonstrating the one-way transmission of laser pulses over 80.1 million km [1].

The first two-way laser transponder demonstration experiment over long distances was carried out between the GGAO and the MESSENGER (MErcury Surface, Space ENvironment, GEochemistry, and Ranging) spacecraft [16]. Again, the transponder demonstration was an experiment of opportunity when the spacecraft was at a distance of 6 million km before an Earth flyby on its 6.6 year journey to the inner planet Mercury. MESSENGER is equipped with a laser altimeter to map out the Mercury surface in a similar fashion as MOLA was used on Mars. This experiment had two lasers working, therefore a complete asynchronous transponder demonstration became feasible. Both terminals fired laser pulses at each other and the spacecraft telemetry transmitted the fire and detection epochs to the ground station, using the stable satellite clock as a constraint. The satellite clock runs on an internal time scale, which is periodically synchronized to UTC and over the duration of the experiment the clock was stable to about 1 ppb. A total of three sessions were performed, where the spacecraft laser was turned on for 5 hours each, while the spacecraft itself scanned the direction toward Earth slowly over an area of 3.2×3.2 mrad at a rate of 16 μrad

s^{-1}. The spacing of the tracks was set to 32 μrad. The transmission of the pulses at GGAO were recorded on event timers as well as the detection of receiver events. GGAO was tied to UTC within 100 ns. A total of 16 pulses were recorded by GGAO on 27 May and 32 pulses on 31 May on the downlink path. The uplink path was more effective, because the ground station did not require to scan the telescope. The evaluation of the range and the rate of change used the equations (5.2) and (5.3). To date, this is the largest distance over which transponder laser ranging has been successfully carried out. The range turned out to be 23964 675433.9 ± 0.2 m with a rate of change of 4154.663 ± 0.144 m [16], which is an excellent achievement.

Apart of these feasibility studies, there are practical applications for transponder activities. One example is the Lunar Reconnaissance Orbiter (LRO) [20]. This satellite also uses a laser altimeter, the Lunar Orbiter Laser Altimeter (LOLA) for the precise mapping of the topography of the Moon, operated between 2009 and 2016. The laser timing is facilitated by an ultra-stable crystal oscillator. Simulations prior to the launch, however, revealed that a drift of the clock frequency may compromise the accuracy of the seleno-location of the measured ground track. The corresponding effect on the actual range measurements in comparison does not suffer from this clock drift in a noteworthy way. The range measurements were designed for a resolution of 10 cm. This led to the decision to adjust the clock and for that matter the ground track by a one-way laser link between Earth and the LRO satellite. The high-gain antenna for the satellite telemetry was fitted with a small optical receive telescope, which consisted of a diameter of 1.9 cm fiber launcher, matched to a bundle of seven optical fibers. The tolerance of the receiver is 30 mrad. The optical telescope is looking through a hole of the antenna dish. While the LOLA ranging hardware is nadir pointing, the antenna can be adjusted dynamically and is usually pointing to Earth with the center of the cone at the White Sands missile range, which acts as the telemetry ground station. The field of view of the optical telescope is large enough to almost cover the entire Earth when the telemetry station is acquired.

The end of the fiber bundle is injected into the detection channel of the LOLA instrument via a beam splitter. A narrow spectral filter of 0.3 nm FWHM removes background light from the Sun illuminated Earth from the ranging channel. There are two gate windows involved. One is 8 ms long and activated before the LOLA laser fires. It provides the 'Earth window' for the one-way range measurements from the ground stations. The second window is 5 ms long and provides the detection window for the primary mapping task of LOLA. The windows have a fixed time dependence on the LOLA laser fire, so that the one-way ranging and the Lidar operation did not interfere with each other. The long-term operation of the clock synchronization process by the transponder operation determined a fairly linear drift of the oven controlled ultra stable quartz oscillator of 3.8×10^{-12} per day over more than 2 years [17].

5.3 Simulations of transponder operations

The active development of transponder experiments requires a close connection to space agencies and satellite programs. This is nothing that can be realized by a tracking station alone. However, this does not exclude a laser ranging ground station

Figure 5.5. Illustration of the transponder testbed arrangement. The transponder (A) and the ranging station (B) are both located in the same spot on the ground. The reflection off from a geodetic satellite with the reflector cross section σ provides a realistic ranging scenario.

Figure 5.6. The transponder demonstrator is mounted on the ranging telescope and aligned with a target several kilometers away (left). Once aligned, the system was used for ranging in the transponder simulation configuration (right).

from exploring this technology. Actual ranging stations provide a unique opportunity to test and explore mission scenarios and transponder prototypes in a realistic context. All the transponder realizations shown in the previous section consisted of ranging systems, which were looking at each other from a large distance. In our local test environment, we have both systems located at the same spot, looking at the same geodetic satellite, which is used as a mirror in order to reflect the measurement signals from one of the two systems to the other and vice versa [14]. Figure 5.5 sketches this situation. The range-dependent signal loss caused by the distance of the satellite, the divergence angle of the ranging system and the reflector cross section σ of the satellite can then be considered as the respective simulation distance that the transponder application has covered. This can be calculated by

$$r_t = r_s^2 \sqrt{\frac{4\pi}{\sigma} \frac{1}{T^{\sec\Theta}}}, \tag{5.4}$$

starting from the laser link equation [4]. In this equation r_s is the instantaneous distance between the ground station and the satellite used as a mirror. σ is the reflector cross section of the remote target. The atmospheric attenuation enters via the transmission T and the zenith distance Θ. This approximation also accounts for the fact that the transponder module is considered to be located in space with very little or without atmosphere. Figure 5.6 depicts the transponder simulation breadboard on the top of the telescope of

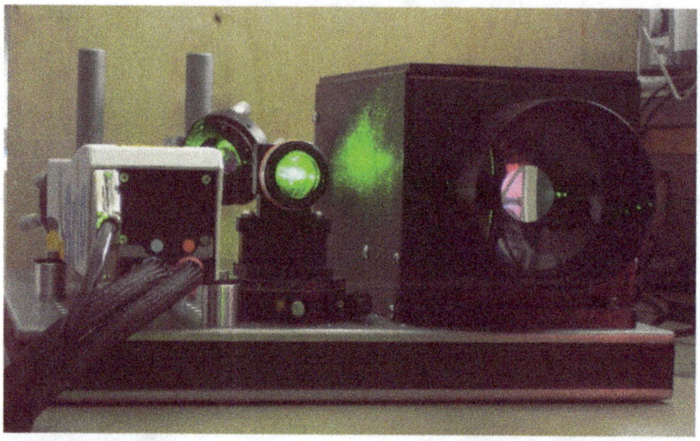

Figure 5.7. The transponder simulator testbed in the laboratory. The laser is mounted on the left side and the receiver section is placed on the right. The transmit optics is illuminated by the laser.

the WLRS. The telescope was pointed to a remote target, several kilometers away. The transponder simulator could then be adjusted to the same landmark, including a small correction for the parallaxes between the two telescopes. Under ideal conditions, the data rate of both terminals should be similar. This would provide the best resolution for the range and range rate measurement. However, that would mean that the laser power and telescope aperture of both systems would be similar, or at least balanced for the same data rate. This applies if both systems are considered to be cruising in space. Then considerable limitations apply since a typical satellite payload faces mass and power consumption restrictions. In cases where one of the terminals is located on Earth, the setup may become considerably asymmetric. There can be a much larger telescope on the ground and the laser can be made much more powerful also. Such a setting can either lead to smaller and lighter payload requirements for the transponder, or increase the achievable measurement distance.

Our testbed demonstrator was constructed following these considerations. It featured a small breadboard system (figure 5.7) with a transmit telescope of 3 cm and 10 cm for the receive telescope. The system was placed on top of the 75 cm diameter optical telescope of the WLRS and was tested under several different configurations and with satellites at different distances [14]. The main observation setup used the WLRS as a transmitter and the demonstrator unit as the detector and vice versa. We also used the main system and the demonstrator as parallel operating ranging systems. In terms of the link budget for three different types of satellite, we obtained table 5.1.

The repetition rate of both lasers were not the same. The WLRS was ranging with a repetition rate of 10 Hz, while the transponder simulator operated at 20 Hz. From table 5.1 we can see that even the small telescope on the simulator testbed provided laser returns from a LAGEOS satellite. The estimated link budget still provided a detection signal level of 1.8 photo-electrons (n_{pe}) in accordance with equation (4.2). The typical transponder setting with signals going from A $\rightarrow$ W and W $\rightarrow$ A is fairly

Table 5.1. Link budget estimates for the transponder test for different satellite targets. The transponder testbed is denoted as (A) and the WLRS ranging system as (W).

Configuration	n_{pe} AJISAI	n_{pe} ENVISAT	n_{pe} LAGEOS
W–A	4.5 k	1.5 k	10
A–W	10 k	3.5 k	24
W–W	240 k	3.5 k	547
A–A	817	272	1.8

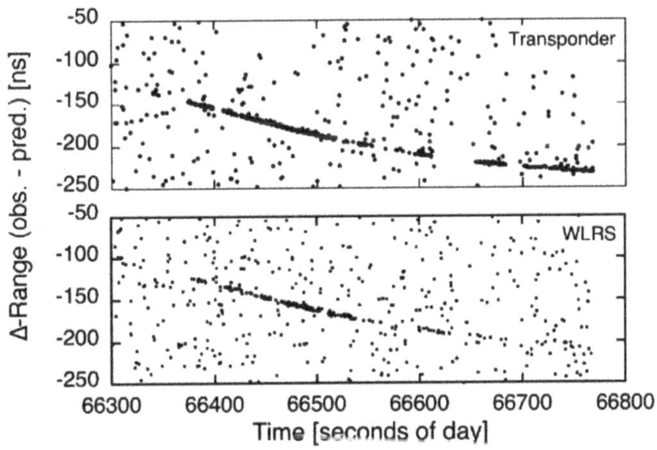

Figure 5.8. Asynchronous transponder operation between the WLRS and the testbed transponder. The two system operated independently on different repetition rates, namely 10 Hz and 20 Hz.

well balanced, with roughly a factor of 2 between them. When we convert the transponder simulation measurement for the satellite LAGEOS to the equivalent interplanetary transponder distance of equation (5.4), we obtain a workable distance of 0.44 astronomical units, corresponding to ≈ **66 million kilometers**. We recall that the detection telescope on the breadboard system only had an aperture of 10 cm and was not optimized for that purpose. Figure 5.8 shows a typical observation example as they were obtained from the two independent ranging systems operated in the transponder configuration.

5.4 Delay compensation and common clocks

The Geodetic Observatory Wettzell operates all of the four techniques of space geodesy. The reference points of all techniques are tied together by a local survey and therefore it is considered a **fundamental station**. There are three VLBI telescopes, two SLR systems, several GNSS receivers and a DORIS beacon. In addition to that the observatory features a time laboratory, which reports UTC(ifag) to the BIPM and therefore contributes to UTC. There is a superconducting gravimeter on the

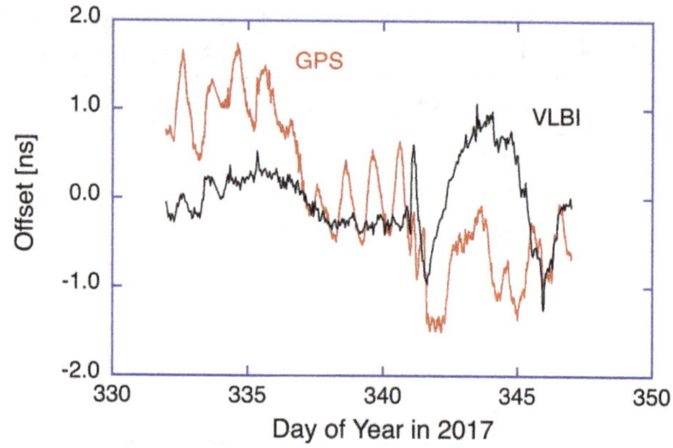

Figure 5.9. Comparison of two remote masers in Wettzell and Matera via GPS and VLBI during the Cont17 campaign. Systematic errors contaminate the clock offsets, therefore we do not see the typical maser behavior in this comparison.

observatory and a large ring laser gyroscope. Other auxiliary measurement systems like a meteorological station, a cloud sensor and a water vapor radiometer are also complementing the geodetic measurements. So far we have seen that all the main measurement techniques in space geodesy are based on the measurement of the time of flight and therefore they are purely geometric in nature. From that point of view it is desirable to utilize a common clock for all the measurement techniques in order to minimize systematic errors and to exploit time as an additional tie. An example illustrates this goal.

During the Cont17 campaign, a two week long continuous VLBI observation session, we had the opportunity to compare the masers in Wettzell and Matera over two simultaneous links, namely via VLBI and via GPS. This is shown in figure 5.9. None of the two techniques reproduces what a direct side by side comparison of two masers would show. We note that the clock in each of the two different measurement techniques absorbs systematic errors of the measurement process. This contamination of the clock offsets by systematic errors is very apparent in figure 5.9. However, that also means that the coordinates of the respective systems are also contaminated by systematic errors. The two-way technique SLR suffers from similar problems, which is illustrated in figure 5.10.

When we take a ps-pulse laser and put a beam splitter into the optical path, so that two detectors would measure the time difference corresponding to the difference in the optical path, there is no variable delay between the two measuring points observable. Since it is not possible to measure the time of flight in the optical domain, we have to introduce photodetectors and a timer for the measurement. The conversion of the optical pulse to an electric signal instantly introduces a delay, which depends on temperature, the signal intensity and subtle ground potential fluctuations between detector and timer. The calibration measurement in SLR is a technical realization of this simple sketched setup. Repeating this calibration over and over again over several days reveals some variability of about 40 ps in a value,

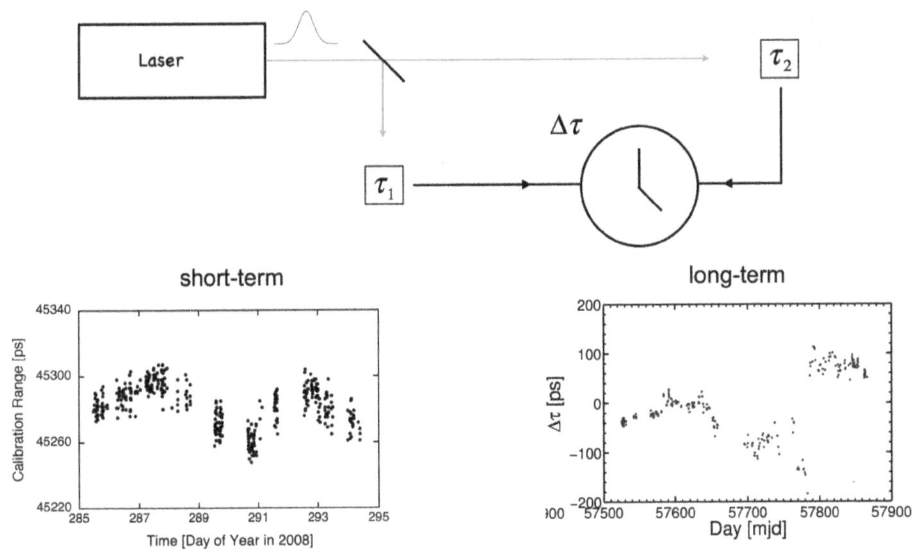

Figure 5.10. Sketch of a simplified optical ranging setup. The pulses from a laser source are split up in two signal paths of different length at a beam splitter (top). Delays occur where the optical signal is converted to the electronic domain, which are sensitive to temperature, floating electrical potential and signal intensity. In the short term the effect is fairly small (bottom left), but may become substantially larger in the long term (bottom right).

which we would hope to be constant. However, over a whole year, there are even much larger excursions visible, increasing the range of variability almost by a factor of 10. This observation does not per se mean that we are dealing with an observation bias of 40 or more ps as time goes on. It means that the system experiences some variability, which the calibration has to take out. The question is, how well can this be achieved. When a ranging system experiences some variability, we can be sure that some of it will not be calibrated out.

The most prominent error source in this context is the intensity-dependent detector delay for avalanche photodiodes and the gate open delay for photomultipliers. Both detector types are leading edge detectors, utilizing a constant threshold. Since the detector rise time is bandwidth limited, these detectors exhibit a nonnegligible response delay between a weak and a strong echo signal from the satellite. Compared to the satellite return, the calibration signal produces a much smaller intensity variation, which is an asymmetry that may lead to a bias in the calibration correction. A similar problem can develop, when the gate-open delay differs between calibration and ranging, at least for a photomultiplier detector. Range gates are very predictable and in order to reduce the influence of background light, the range gates are short, typically between 200 ns and 1 μs. The calibration gate window, however, can be fairly large, if the delay between the laser fire command and the actual firing of the laser is not well determined. A photomultiplier loses the plate charge as time goes on, which in turn increases the response delay. The electrons are not accelerated as much at a lower voltage and therefore take longer until they reach the anode. In extreme cases of a calibration gate with a response delay of more than 100 μs, additional delays of up to 120 ps have been observed with a laser utilizing a dye Q-switch.

5.4.1 The lossless distribution of time and frequency

A way around this problem is obtained by introducing delay compensation and performing closure measurements [7, 13]. Delay compensation can only be achieved in the optical domain, because it requires a completely reciprocal two-way signal propagation approach. We have briefly discussed that in section 2.6.2. Figure 5.11 recalls the simplified block diagram, which is sketched like an unequal arm Michelson interferometer. The signal source is a mode-locked fs-pulse laser. A hydrogen maser delivers the 100 MHz locking frequency. This combination has the property that it provides ultra-low noise frequencies, both in the optical and microwave regime. Furthermore, for that reason there is also very low jitter in the pulse separation, a very stable pulse intensity and the pulse width is shorter than 100 fs. This laser output is directed onto a beam splitter, where a small amount of the signal is taken out of the beam and guided over a short distance to a mirror where it is back reflected. The main pulse train is passing over an adjustable delay line and then through a mono-mode fiber of up to 700 m of length to the location of application. This can be the laser ranging facility or one of the VLBI telescopes, the timing system of Wettzell, the GNSS receiver platform, the ring laser gyroscope or the gravimeter laboratory. At the end of the fiber there is another mirror, which returns most of the signal back toward the laser source. The remaining intensity of the pulse train is detected by a photodiode and the repetition rate is converted back

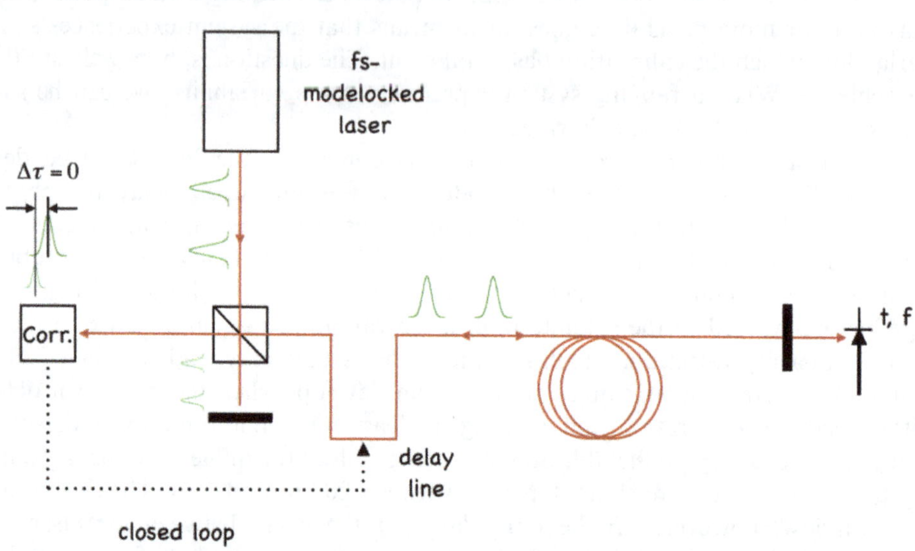

Figure 5.11. Block diagram of the two-way optical signal delay compensation system, illustrated in the shape of an unequal arm Michelson interferometer for fs-pulses from a mode-locked laser.

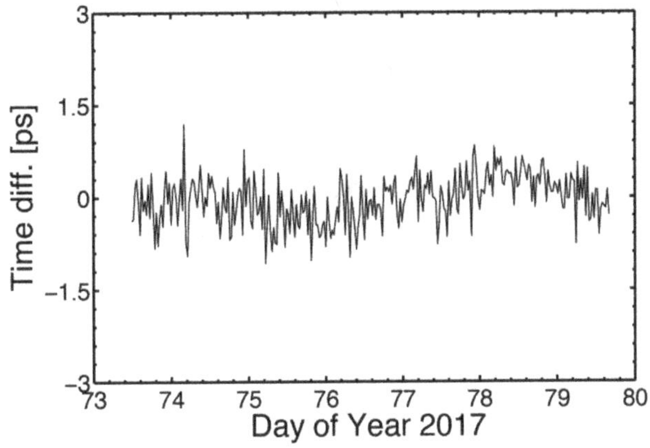

Figure 5.12. Example of an observed time series of the variation of the delay between the mode-locked laser and on of the outputs several hundred meter away. The variation stays well within ±1 ps.

to a clock frequency. In addition to that the fs-pulses are also used to generate a time scale, deriving the PPS signal and subdivisions of that down to 10 ns. Near the laser source, the returned pulses from the long arm and those from the short arm are super-imposed in a correlator arrangement. We recall that the pulses are very short, less than 100 fs. Normally one would expect that the sharp pulses disintegrate over the long distance in the fiber. However, the fiber is composed out of two parts, a typical mono-mode fiber and a conjugate fiber. This ensures that the pulses at the beginning and the end of the long signal fiber are as sharp as they are coming out of the mode-locked laser. Within the correlator unit it is evaluated by how much the pulse from the long arm has shifted relative to the one of the short arm and the delay line is driven such that the acquired delay is actively compensated. The result of this effort is a stable delay value between the source and the end of the fiber line.

Figure 5.12 provides an example of the achieved delay stability between the source signal from the mode-locked laser and the end of the signal transmission line. This time and frequency distribution system does not only feature one actively stabilized time and frequency line, but up to fourteen independent links, originating from the same laser source.

In practice it means that it is now possible to operate an entire fundamental station on a single common clock, compensating local drift effects, induced by varying temperature, impedance changes and strain on the signal distribution lines, caused for example by moving telescopes. Furthermore, due to the fact that we are dealing here with broadband optical signals, the phase of the clock frequencies is much better defined. They are no longer depending on trigger threshold settings in the presence of electrical potential differences. How large are these variable line delays anyway? Figure 5.13 provides the error signal as a function of time, which drives the variable delay line for two different examples. The left diagram was taken for a short fiber line of about 20 m inside a temperature controlled laboratory. We can clearly see the de-icing cycles of the air conditioning system, causing variability

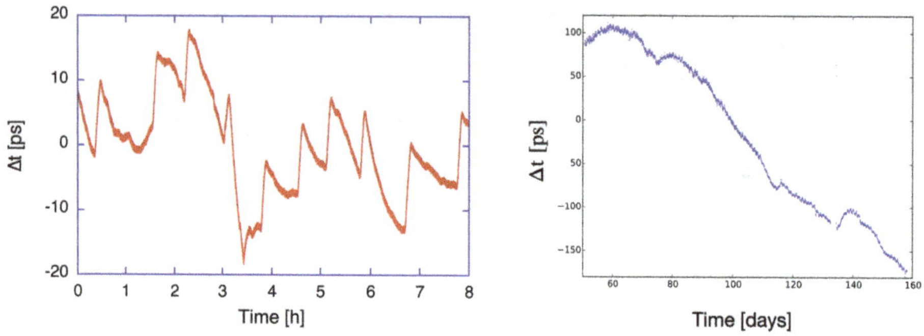

Figure 5.13. Example of an observed time series of the variation of the delay between the mode-locked laser and on of the outputs several hundred meter away. The variation stays well within ±1 ps.

of 30–40 ps over a total of 8 h. The diagram on the right side looks at a long fiber line of about 450 m in an underground cable tunnel between buildings over about 100 days. We can see a drift of more than 250 ps, mostly caused by a seasonal change. The daily temperature changes are also visible in this dataset, but much smaller in comparison.

5.4.2 Two-way optical time transfer

The system described in the previous section transfers time and frequency reliably and with high stability without any other external signal requirements. It distributes the frequency of a maser source with a constant transport delay. By assigning a particular epoch to one of the laser pulses, it can distribute a time scale with very high accuracy, once the exact delay for each fiber line has been established. It also supplies more than one other location with a delay compensated clock signal. In this section here we look at another method, which is suitable for a direct clock comparison between two locations [6]. However, we need to keep in mind that it compares two locations, but it does not transfer time and frequency from one spot to another. With all this said, the great advantage of this system is a simple setup and the operational flexibility [8]. The most suitable application of this concept is in a local observatory environment. For other more widely spaced comparisons, there are further technical possibilities, which do not apply to the optical regime. They are discussed in [9] and mentioned here for completeness. The functional layout is sketched up in figure 5.14. The basic concept of operation is realized either over rf cables or over optical fiber lines and can be described in the following way. Each of the two ends of the comparison line is terminated with identical instruments A and B, each of which contain an event timer (ET) and a pulse generator (PG). The two instruments exchange signals either by a coaxial cable or an optical fiber line. The measurement process starts by system A emitting a pulse, the arrival time τ_{A1} and τ_{B1} of which are timed on the event timer in A and in B. This corresponds to the upper part of figure 5.14. Then the roles of the two systems are exchanged and we observe

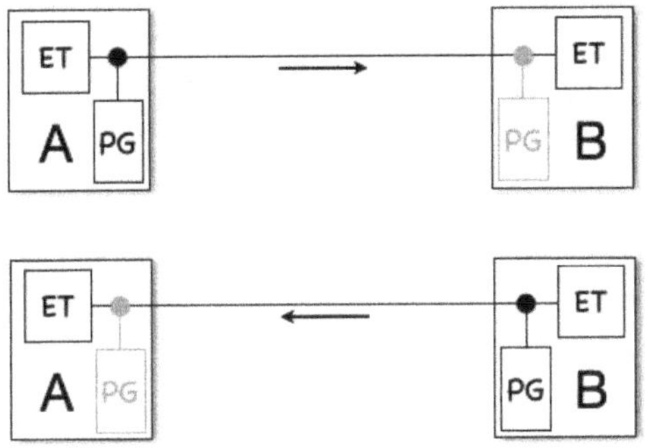

Figure 5.14. Example of an observed time series of the variation of the delay between the mode-locked laser and on of the outputs several hundred meter away. The variation stays well within ±1 ps.

τ_{A2} and τ_{B2} as shown in the lower part of the sketch. The difference in the time scale Δ of each unit and the delay δt introduced by the connecting line is obtained by

$$\Delta = \frac{1}{2}((\tau_{B1} - \tau_{A1}) + (\tau_{B2} - \tau_{A2})) \tag{5.5}$$

and

$$\delta t = \frac{1}{2}((\tau_{B1} - \tau_{A1}) - (\tau_{B2} - \tau_{A2})). \tag{5.6}$$

Repeating this measurement over and over again, reveals the stability of the offset between the two devices. For completeness we need to add that each terminal of the TWOTT link requires an input of 100 MHz clock frequency and the PPS signal. From that point of view it is not suitable for the actual transfer of time and frequency. However, this system is well suited to compare two remote clocks with moderate hardware requirements.

5.4.3 Closure measurements

We have seen now clearly that all of our geodetic measurement devices are subject to systematic delays in the measurement process. Since these measurement techniques realize all time-of-flight measurement concepts, these systematic signal delays are causing a bias and these offsets will not be constant. A way out of this dilemma is provided by the concept of closure measurements. A closure means, we are working our way around in a circle from one piece of the measurement chain to the next. If everything is stable and all delays are captured in one way or another, the closure yields an offset of zero and that is desired quantity. If, however, the offset is not zero, we have come across an uncorrected bias. Figure 5.15 illustrates this concept graphically. The delay compensated distribution of clock pips along with the 100 MHz clock frequency allows a clock comparison with the resolution of ± 1 ps as demonstrated in figure 5.12 and in [7]. We have used the TWOTT technology to compare the phase of the reference frequency between two telescopes at the

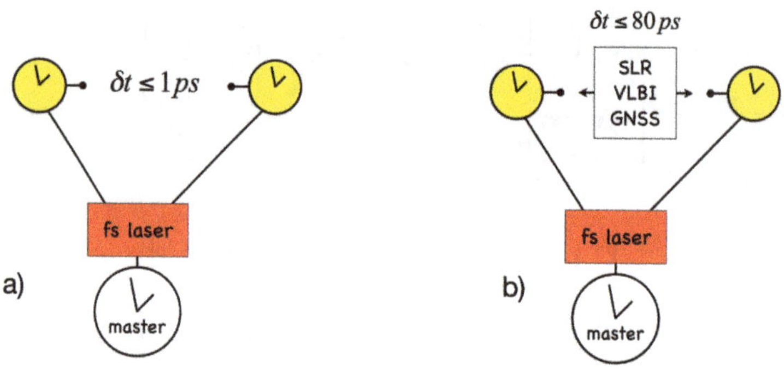

Figure 5.15. Illustration of the closure measurement concept, realized with the time and frequency distribution system treated in section 5.4.1. Two branches of the delay compensation agree to within ±1 ps (a). It means that we can identify systematics by measuring in closure around a full measurement system (b).

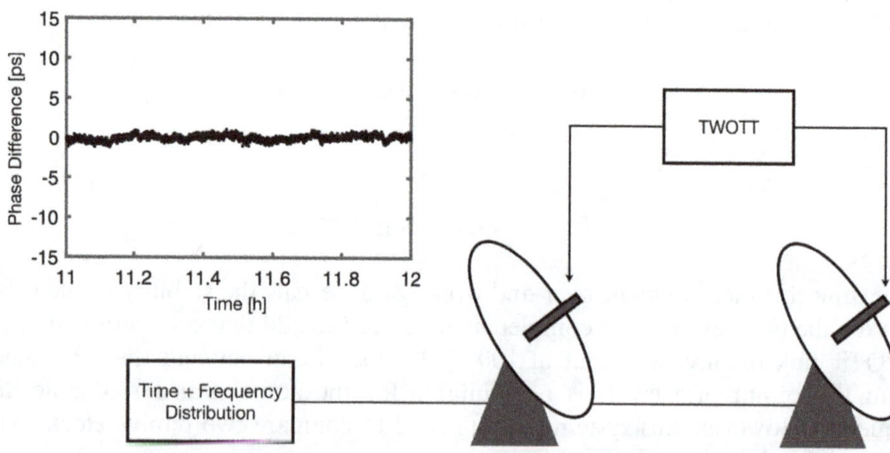

Figure 5.16. A common clock comparison between two radio telescopes connected to the common time and frequency distribution system. The phase difference between the 100 MHz reference frequencies in the feedhorn of the two telescopes stayed constant with a standard deviation of $\sigma = 0.34$ ps.

observatory Wettzell, which were connected to a common clock via the time and frequency distribution system. The result is shown in figure 5.16. The two telescopes are about 100 m apart and the TWOTT system was probing the relative phase difference of the provided reference frequency at each feedhorn. Over one hour there was no noteworthy shift in the phase relationship observed. The standard deviation of the relative phase difference between the two telescopes was calculated as $\sigma = 0.34$ ps. This is an excellent result, because the telescopes were slewing through more than 50 radio sources during the time of this frequency comparison. For the evaluation of the actual baseline length of these two telescopes, we had taken the local survey as a constant and subtracted this from the measured baseline. Due to the common clock operation, there is no additional clock correction required. The remaining residuals are presented in figure 5.17. Apart from a slow upward drift, we

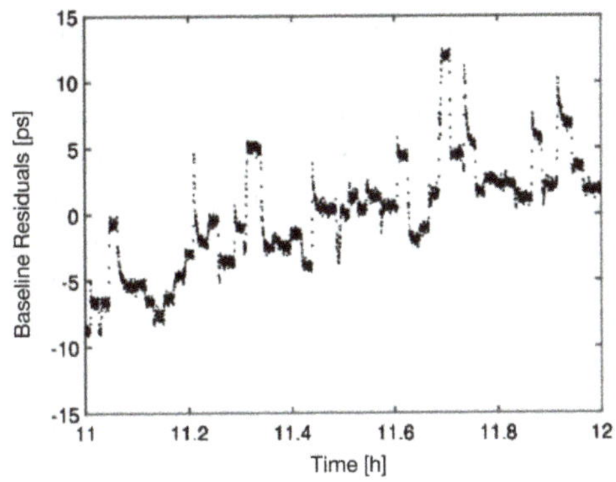

Figure 5.17. A one hour measurement sequence between two local VLBI telescopes, operated on a common clock with a delay compensated time and frequency distribution. Apart from a apparently temperature related drift, we also see strain effects from the slewing of the telescope.

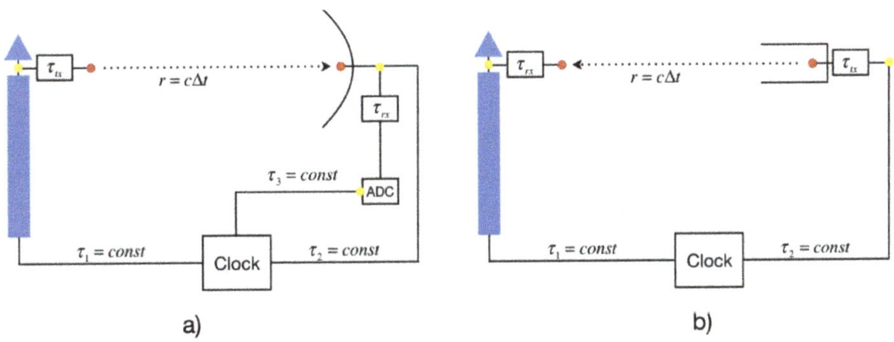

Figure 5.18. The closure measurement concept utilized with the stable delay compensation, illustrated for VLBI and SLR operations. By going from the clock to the fiducial reference target to the measurement system and from there back to the clock, all delays should cancel in a complete measurement circle. The concept for VLBI is shown in (a) and for SLR in (b).

can see several jumps of up to 10 ps in the remaining residuals. The obvious drift is probably related to temperature variations, somewhere in the receiver chain. The jumps however, are certainly not related to temperature. Some of the jumps are also showing signs of relaxation. It is very likely that we see the effects of cable strain inside the cable twister of the VLBI telescopes. While the distribution of the clock frequency is fully delay compensated, there is no phase shift on the clock frequency. The receiver signal line is not compensated, therefore, the cable strain is adding to the noise and the systematic errors of the VLBI measurements.

Without the delay compensated time and frequency distribution system this kind of systematic measurement error is inaccessible. To illustrate how a closure measurement can be realized for a VLBI or a SLR system, there is a conceptual layout provided in figure 5.18.

The central starting point for both techniques is the common clock. There is a link of the time and frequency distribution to the central common reference target. The actual delay is known, because the roundtrip delay has been measured with a portable two-way fiber link device, hooked up to the same clock. The microwave signal (a) is radiated in the form of a sine wave from the common reference target, which takes the position of a geometric fiducial point of the observatory. It is detected by the VLBI system and compared to the local phase calibration reference signal generated in the VLBI system, utilizing another two-way link of the time and frequency distribution system. A third delay compensated link connects the digitizer of the VLBI system to the clock. The three clock links have three constant delay values, namely τ_1, τ_2 and τ_3. All three of them are established with the help of a portable delay compensated link in the same way as described above. This leaves τ_{rx} as the only potentially variable delay in the complete closure circle, which now becomes accessible. τ_{tx} is separately controlled internally in the transmitter unit.

The closure measurement technique is simpler to realize in the SLR technique, because SLR is already a two-way measurement technique. Figure 5.18(b) illustrates the process. Again, we start off from the clock. The two-way delay compensated time and frequency distribution connects the clock with the SLR system. The SLR system performs one-way and two-way ranging to the station fiducial point. The arrival of the measurement signal is timed on an event timer on the common target, which in turn is connected back to the clock via another delay compensated clock link, thus closing the circle. Figure 5.19 provides results from an extended closure measurement campaign. There were two SLR systems of the Geodetic observatory Wettzell involved, namely the WLRS and the SOS-W. Both systems exhibit a stable performance over the approximately 200 days of the campaign. The scatter of all of the calibration measurements is of the order of 5 mm. The solid line shows the geometric reference point position, established from several surveys of the common target and the system reference points from both laser ranging systems. The established system closure for both SLR systems falls short by 3 mm or 9 ps. The cause for this bias has not yet been understood. It is not even clear if the issue originates in the ground survey or in the SLR systems. Small errors at this level are extremely hard to diagnose. Nevertheless,

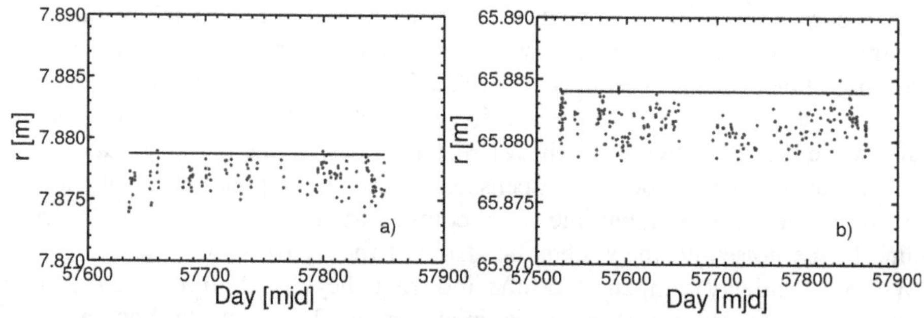

Figure 5.19. Display of the result of a calibration campaign, utilizing the common clock station fiducial WTCO. The measurements relate to the WLRS ranging system (a) and the SOS-W ranging system (b). With the solid line providing the geometric reference point established from several ground surveys, we note that both systems measure stable, but short by 3 mm.

we have to realize that this is a potential systematic error, which reduces the desired accuracy of the measurement system.

5.5 Time transfer over optical laser links

In section 2.5 we have looked at the Einstein synchronization as a viable concept to compare two remote clocks at different gravitational potential and which may also be moving with respect to each other. Furthermore, we have seen that the SLR measurement technique is the physical realization of the Einstein synchronization, provided there is a detector and timer at the satellite target. We recall that the reading of the remote clock t_2 corresponds to the reading of the local clock t_1 + half of the roundtrip time of the optical probe signal, as shown in equation (5.7)

$$t_2 = t_1 + \mu(t_3 - t_1) \quad \text{and} \quad \mu = \frac{1}{2}. \tag{5.7}$$

Repeating the measurements frequently will provide insight in how the two time scales develop with respect to each other. This includes the effect of satellite motion and gravitational redshift. Historically, clock comparisons were made by putting the clocks of interest side by side. With novel techniques of fiber based delay compensation, it is also possible to compare clocks, which are spaced apart by more than 1000 km. Since we compare the tick rate between clocks, we need to realize a delay compensated frequency comparison, as introduced in section 2.6.1. At this point in time this technique is applicable within continents, but not between continents, because such a comparison requires a set of optical amplifiers every 100 km or so. These amplifiers are a bit delicate and must be accessible for service, so they cannot be on intercontinental underwater cables. The alternative is provided by time transfer utilizing a satellite link. It is important to note that we need to go from frequency to time, when an optical satellite link is used. When we do this, we also require a delay compensation system in order to preserve the accuracy of the time comparison. This was briefly introduced in section 2.6.2 and in more detail in section 5.4.1. A satellite link provides the opportunity to compare high performing clocks on a global scale. Furthermore, it is possible to perform tests of fundamental physics with an orbiting clock. Such a clock comparison requires a continuous path from the phase or the time ticks of one clock via the satellite to the target clock.

5.5.1 Time transfer via passive satellite bodies

We have tested this general concept by transferring time from one SLR station to another via diffuse reflections from satellite bodies [10]. Figure 5.20 illustrates the measurement concept in form of a range–time diagram. The distance is shown on the ordinate, while the x-axis marks the time. The top track symbolizes the motion of the rocket body, simultaneously visible by the two stations. The orbit of the satellite can be computed by regularly updated ephemeris. The two time scales refer to the two involved ranging stations. Laser station A, in this case the WLRS fires at the epoch t_{A1} and detects the diffuse reflection from the rocket body at t_{A2}. The laser pulse hits the satellite body at the epoch t_s, in the time scale A. We can compute t_s as

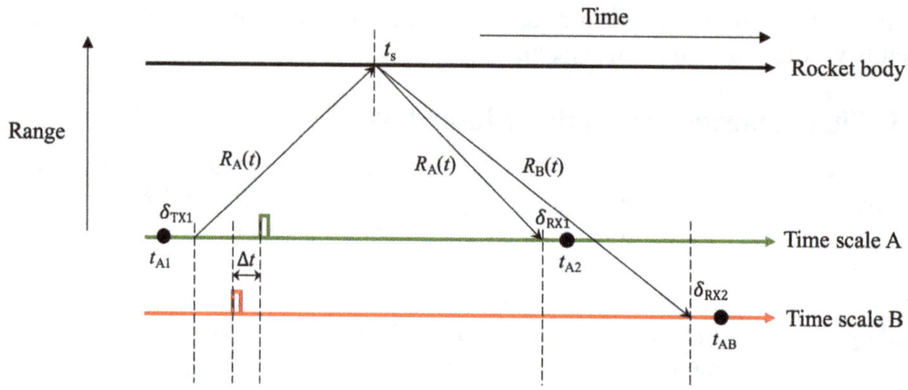

Figure 5.20. Illustration of the basic time transfer concept via a passive satellite target. In this range–time diagram, distance is given on the ordinate, while time progresses along the x-axis. We use diffuse reflections from a satellite body to transmit short laser pulses from one station to the next.

$$t_s = \frac{1}{2}(t_{A2} - t_{A1}) + t_{A1}. \tag{5.8}$$

The offset between time scale A and B is then computed as Δt and we get

$$\Delta t = \frac{(R_A + R_B)}{c} - (t_{AB} - t_{A1}). \tag{5.9}$$

The value of Δt is indicated in figure 5.20 as the difference of the PPS markers in the time scale A and B. We note that the value for $R_B(t)$ is computed from the ephemeris, corrected to the actually measured orbit from the tracking station A. Furthermore, it is important to note that the instant of transmission is timed before the laser pulse passes the reference point of the telescope. Therefore, we have to determine the uplink delay and the downlink delay of each SLR system separately and with the best available resolution, preferably down to a single ps. Utilizing large targets of space debris, such as rocket bodies, provides comfortable ranging signal levels, but has the disadvantage that the location of reflection is not well determined on the target. In addition to that we face the problem that abandoned rocket bodies carry out uncontrolled tumbling motions, which we have to detect and disentangle from our range measurements.

Figure 5.21 provides an example of the ranging data (a) and the light curve for the same pass (b). While the former corresponds to the range measurement of a large slowly changing shape of the target, the latter stems from the scattered sunlight, illuminating the rocket body. The signal corresponds to the current proportional to the reflection brightness, measured in intervals of 10 ms. It is important to note that the two signal sources have a very different geometric dependence. This is illustrated in figure 5.22 and more details can be found in [10].

The orientation of the rocket body viewed from the Earth has some ambiguities, so that the range information is not enough to resolve the tumbling motion. Therefore, it was necessary to also include photometry (light curves) in the measurements. The Sun

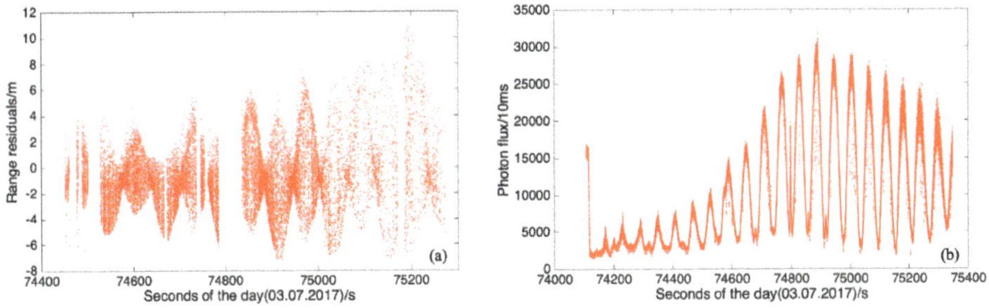

Figure 5.21. Example of the measured track of a tumbling rocket body (a). When the Sun illuminates the rocket body, the part of the signal scattered to the ground station can be detected by a photomultiplier as a time resolved light curve (b). The combination of these two diagrams allows the reconstruction of the rocket body motion.

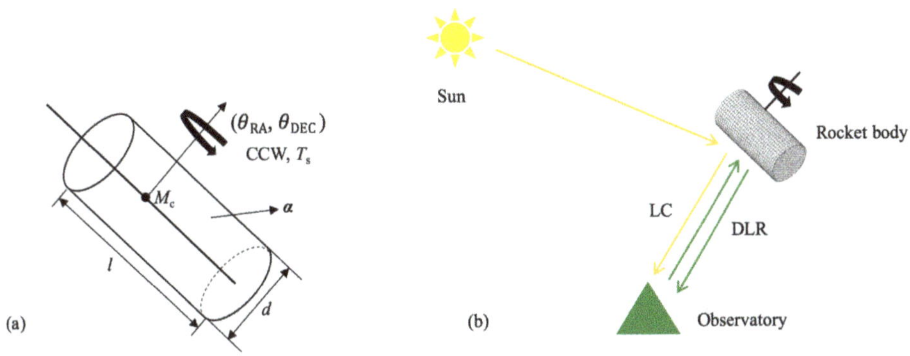

Figure 5.22. The specific model of a tumbling rocket body (a) includes the dimension (length l and diameter d), the center of mass position M_c, reflection parameters α, the tumbling axis orientation (θ_{RA}, θ_{Dec}), the sidereal tumbling period T_S and the tumbling direction (CW or CCW). The right side diagram illustrates the observation vectors for the laser ranging and the photometry.

illuminates the rocket body and some of the light is scattered toward the ranging station. One can see that the range information for the example shown in figure 5.21(a) is very different from the photometric measurements in figure 5.21(b). The combination of the two observables constrains the tumbling motion sufficiently enough to remove this ambiguity. However, what we cannot remove is the large scatter that comes from the illumination of a rocket body with approximately 2.5 m diameter. Figure 5.23 shows the final result. Once the tumbling motion has been removed from the measurements both in Wettzell and Graz, we find an offset between time scale A and B of 2.023 μs ± 3 ns, which is quite remarkable, given the poor reflection geometry. We turn now to time transfer with support of an accurate onboard satellite clock.

5.5.2 T2L2: time transfer by laser link on JASON 2

We recall that the isotropy of the speed of light is a mandatory requirement for optical time transfer. To add the time transfer functionality to SLR we need an additional

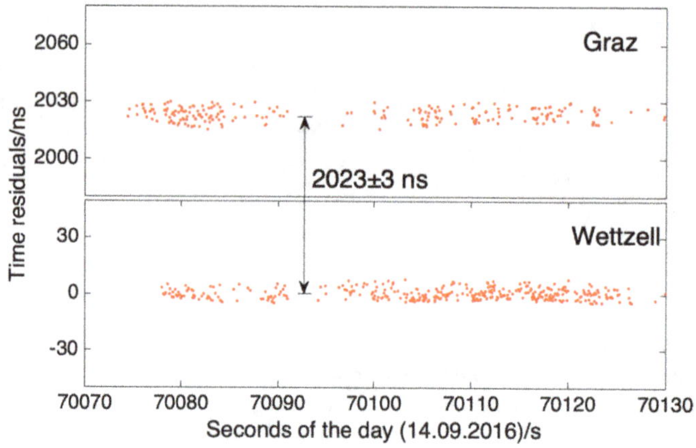

Figure 5.23. Determination of the time scale offset between the WLRS and the SLR station in Graz. After modeling the tumbling motion an offset of 2.023 μs ± 3 ns is obtained.

detector on the spacecraft. It must be very close to the cube-corner reflector (CCR), which acts as the zero-delay reflector that the Einstein synchronization requires. This additional detector provides the laser pulse arrival trigger that is timed on the satellite clock. To properly transfer the time, the ground station must time the laser fire event with the highest possible resolution, preferably 1 ps. Furthermore, we need to separate the SLR system calibration delay into the uplink and the downlink component. Last but not least, it is also important to establish and control the detection delay between the spacecraft detector and the satellite clock. When the spacecraft detector and the CCR are not in close proximity, we have the need for an attitude model, which keeps track of the geometrical differences as the laser signal comes from different directions.

The first application for the evaluation of the optical time transfer was the French Time Transfer by Laser Link (T2L2) package, placed on the JASON-2 satellite [5]. Time transfer in this context means that the ground station clock is compared to the satellite clock repeatedly in order to study the behavior of the satellite clock relative to the time scale of one or several ground clocks. It can also be used to compare two ground clocks via the satellite link. JASON-2 features an ultra-stable quartz oscillator, which is tightly temperature controlled. This is not ideal, but it is what was available at the time of the experiment. A simplified functional concept of the T2L2 payload is presented in figure 5.24 and a detailed report can be found in [19].

Laser time transfer requires a very large field of view for the detector, since the laser beam comes in over a wide range of incidence angles as the payload passes over the laser station. To obtain a fast rise time and low jitter, an avalanche photo diode (APD) in Geiger mode was used. This diode provides a single-photon detection sensitivity and that is in stark contrast to the very high background light level caused by the Earth albedo. To accommodate for a highly variable background light level, another photo diode, shown at the top of the sketch in figure 5.24, constantly detects the instantaneous noise level and the occurrence of a laser pulse. From the recorded noise level, it generates an appropriate detection threshold value and from the

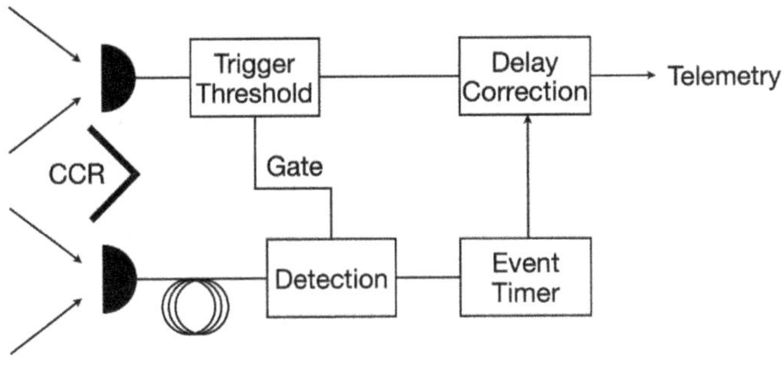

Figure 5.24. The T2L2 payload consists of two detectors, a signal conditioning and gating circuit. Provisions for threshold settings under variable background noise and intensity corrections are augmenting the event timer. The cube-corner reflector array is indicated as CCR.

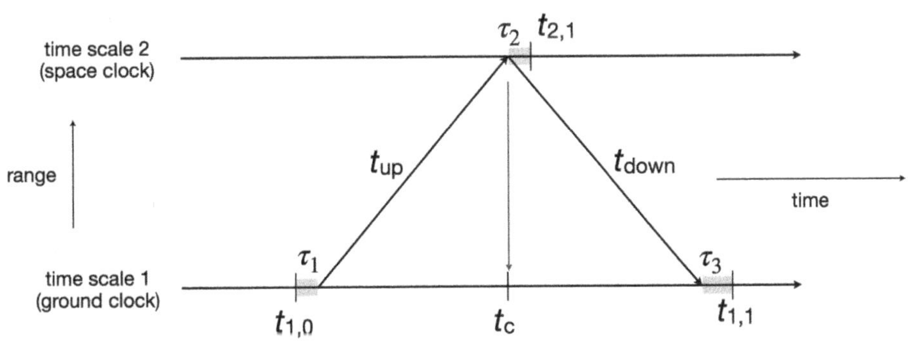

Figure 5.25. Clock comparison between a ground clock and a satellite clock.

detected laser pulse intensity it generates an arming gate signal for the APD, as well as a shot-by-shot time walk correction value. This takes care of the systematic effect that the response time of an APD is intensity dependent. Higher signal intensities are seeding more photoelectrons, resulting in a shorter avalanche build up time. If not mitigated, this variable response delay may well exceed a value of 1 ns, which is unacceptable for the purpose of accurate time transfer.

To allow for the evaluation of the trigger threshold and the generation of the arming gate, there is an optical delay line in the form of a fiber coil, which delays the detection of the time transfer pulse by a constant value. This value is subsequently removed from the measurements numerically together with the value for the time walk correction, computed from the digitized waveform in the analog channel. A full characterization of the T2L2 flight instrument and its calibration is discussed in [12]. It is now time to review the time transfer between a ground clock and a satellite clock in more detail. For this matter we take another look at the time–range diagram of this clock comparison process, which is provided in figure 5.25. The probe signal (laser pulse) takes the time t_{up} to propagate between the reference point of the ground station up to the reflector, detector pair at the satellite.

$$\frac{r}{c} = \frac{1}{2}(t_{1,2} - t_{1,0}) = t_{up} \tag{5.10}$$

The instant where we initialize the clock comparison is $t_{1,0}$. This is the moment when the clock reading is latched by the laser start detector and this is marked with a vertical line in the ground clock time scale 1. From there it takes the time τ_1 until the laser pulse passes through the system and reaches the system reference point. This is indicated by the gray shaded area near $t_{1,0}$.

Then it takes the time t_{up} from equation (5.10) to reach the reflector and detector ensemble at the spacecraft[1]. There is another delay τ_2 between the arrival of the laser pulse at the detector and the time latched by the onboard clock. For low earth orbiters, like JASON-2, we can assume that the uplink and the downlink time is equal, which explains the value 1/2 in the equation. When the ground clock detects the return pulse, the measurement process adds the third delay value τ_3 to the measurement. In order to arrive at the value t_c, which is physically the instant when the laser pulse has reached the reflector at the satellite, we have to remove all the delays from the time transfer measurement. None of these delays can be expected to remain constant over time. Therefore, it is most important to make the delay at the satellite small and as constant as possible. Furthermore, it is desirable to include a suitable calibration capability, so that this delay can be maintained as good as possible. The ground station delays also need to be calibrated on a regular basis, separate for the uplink and the downlink part. Finally, we also require the clock resolution for the time tagging to operate at the ps level. Otherwise we would lose resolution on the time comparison unnecessarily.

The single-shot resolution from the T2L2 operation turned out to be around 80 ps, which averaged down to 7 ps after 30 s of time transfer operation with the WLRS laser operating at a repetition rate of 10 Hz. Longer integration times did not improve the resolution, because the satellite clock was not stable enough [11]. The WLRS was referenced to a maser. The correct reference between the Wettzell time scale and the event timer epoch setting turned out to be a significant problem in practical operations. We have observed considerable jumps of several ns between consecutive synchronizations from one day to the next.

5.5.3 ELT: European Laser Timing using the ACES clocks on the ISS

The European Laser Timing (ELT) is a follow-up experiment on T2L2. The ACES payload is located at the Columbus module of the International Space Station (ISS). It was launched on 21 April 2025 on a Falcon 9 rocket under the mission identifier CRS SpX-32. Figure 5.26 is an artists impression of the deployment (right side) and an inside view on the ACES hardware. The ACES payload [2] contains two different clocks, an active hydrogen maser (SHM) and the cold atom cesium clock (PHARAO). Furthermore, the payload features two independent clock comparison techniques, namely the microwave link (MWL) and the optical time transfer link (ELT). There is a

[1] For the sake of simplicity we assume the refraction correction as correctly established and applied.

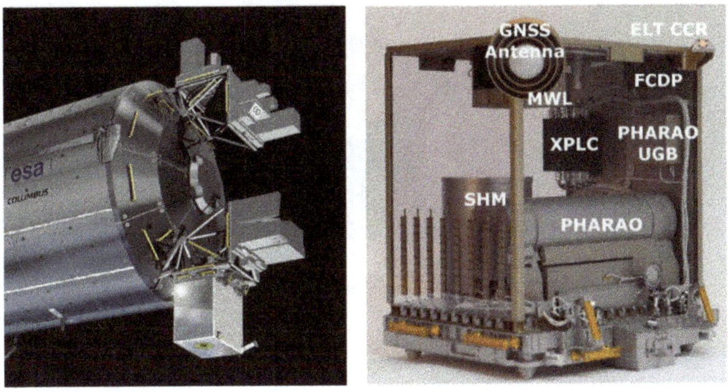

Figure 5.26. An artist impression of the deployment of the ACES payload at the Columbus module of the ISS (left). The inside view depicts the maser (SHM), the cesium cold atom clock (PHARAO), the microwave two-way link (MWL), the GNSS antenna and the ELT package with the CCR reflector on the top right. (Image credit: ESA.)

GNSS receiver onboard that can track the path of the clocks in space and there is the CCR in close proximity to the optical time transfer detector unit. In order to comply with the limited electrical power and weight resources, ELT was designed to minimize the complexity in space and shift it toward the ground station. Apart from the clocks, the ELT package differs from T2L2 by only utilizing one single-photon avalanche diode, actually the same type as it was used on T2L2. The APD is gated on from the ISS time scale every 10 ms. That means that the ground station has to actively control the laser fire instant, such that the laser pulse arrives at the detector within the first 100 ns after gate on. This is a very dynamic requirement, because the range changes quickly and permanently. Furthermore, there is no adjustment for the threshold setting of the detector available and there is no correction for the detector time walk either. The laser pulses must be adjusted by the ground station to the point that single-photon detection is ensured both at the ISS and the ground station during the measurements. ELT is using the timer functionality of the MWL. It does not require an event timer of its own. Instead of the 9.5 kg payload weight and 45 watts of power consumption of the T2L2 experiment, ELT features less than 1 kg in weight (including the CCR array) and much less than 1 watt of power consumption. This made it possible to include ELT into the ACES experiment.

Let us take a more detailed look at the required properties of the detector and the SLR system to provide highly accurate clock comparisons. In order to detect the SLR pulses over a wide angular range, the ACES detector features a angular acceptance range of 120°. The incoming light hits a diffuser window and produces diffuse isotropic radiation. Some distance away from the diffusor plate, there is a baffle that only allows a small light cone to progress further. A spectral filter with the passband centered around the laser frequency at $\lambda = 0.532$ nm, reduces the amount of background light significantly. Another baffle reduces the light cone even further. The remainder of the light eventually illuminates the active area of the photodiode after passing through a neutral density attenuation filter. Since the nadir pointing

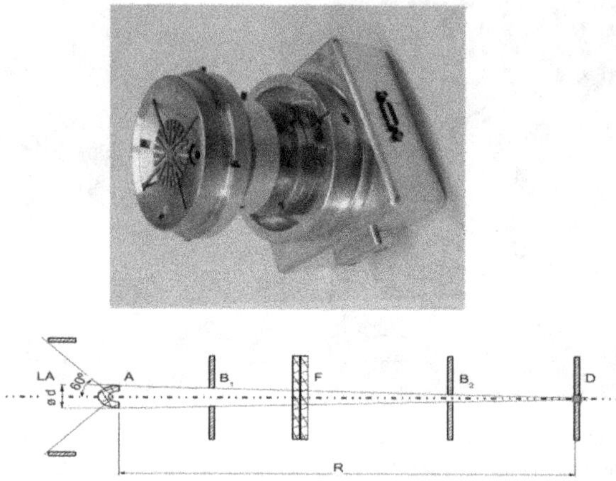

Figure 5.27. The Engineering module of the ELT detector (top) and the optical layout (bottom). The beam block in the front of the detector blocks out the incoming strong background light from the nadir direction. The cutouts produce an angle dependent attenuation, so that the detected light level is fairly constant over the entire range of acceptance angles. (Photo credit: I Prochazka)

detector has the most amount of background light coming from that direction, there was a beam block added, which blocks the light from the nadir direction. This beam block is structured such that it takes out most of the angular intensity dependence of the incoming beam. The total attenuation of this detector light path is 6 orders of magnitude. Figure 5.27 presents the engineering module of the detector and also provides the optical setup. We need to operate at the level of single-photon detection to avoid the time walk bias. Therefore, it is important to note that the attenuation of the detector, the uplink divergence loss ($I \propto R^{-2}$) and the round trip divergence loss ($I \propto R^{-4}$) have to be balanced such that the space detector and the ground detector, both operate safely in the single-photon regime.

SLR is a two-way measurement technique, therefore much of the clock drift cancels out in the measurement process. The advantage is the long-term stability of this measurement concept. We have performed an evaluation experiment [15] with the WLRS operated in the transponder simulation mode (see section 5.3) in order to study the properties that the optical time transfer would add to the ACES project. T2L2 performed very well, but was limited by the stability of the USO clock. Figure 5.28 presents an Allan deviation diagram of the sensitivity of the WLRS for high-resolution time transfer. Integrating range measurements to ACES over 100 s provides a resolution of 0.2 ps for the time transfer. The two-way ranging provides an inherently stable operation scenario, with a constant stability of 9 ps over more than 10 days. ACES is suitable both for common view and non-common view time transfer; the latter enables tests of fundamental physics, by studying a fast moving clock in a low earth orbit. Common view time transfer to the ELT package with two or more SLR stations automatically would provide the possibility to compare two very stable optical clocks on ground via the ELT time transfer link. This is known in the literature as chronometric leveling.

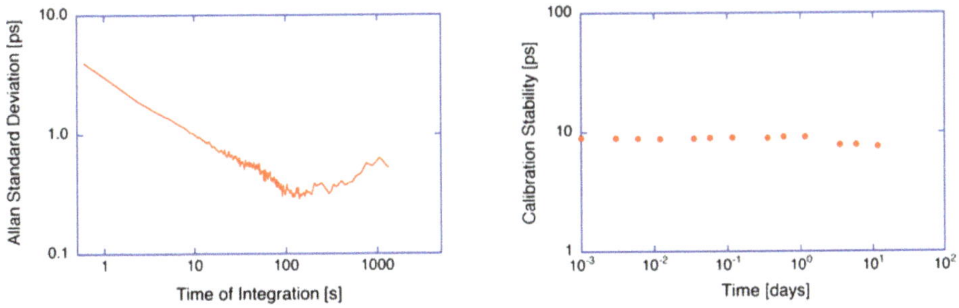

Figure 5.28. The short-term sensitivity and the long-term stability of the WLRS evaluated by the evaluation experiment. After 100 s of averaging, a resolution of 0.2 ps is reached (left). For a large number of measurements, the stability remains at a level of less 10 ps over more than 10 days (right).

5.5.4 Chronometric leveling with the ACES package

The tick rate of a clock, actually any clock, depends on the physical properties of the location of operation, whether it is a clock exploiting the energy state of an atom or an hourglass. When they are compared at different gravitational potentials or under different velocities, they behave in the same way. So let us consider two identical optical clocks, which are operated side by side at the same gravitational potential and we find that they agree perfectly. This means the frequency of the interrogated clock state of both clocks is identical. Now we take one of the clocks and put it somewhere else on a mountain.

From general relativity we expect that the clock farther away from the center of gravity will run faster than the other clock closer to the center of gravity. In section 2.5 we saw that the clocks will differ in the tick rate by a correction value proportional to $(1 + gh/c^2)$. A change of the tick rate of 1 part in 10^{18} of a clock on the Earth surface corresponds to lifting it up by 1 cm. Only optical clocks are good enough to be used for this purpose. In order to exploit this effect experimentally, we are preparing the experiment sketched in figure 5.29. The main goal is the comparison of two remote strontium lattice clock, one of them located at the Geodetic Observatory Wettzell and the other operating at the German NMI in Braunschweig. To cover the distance, we exploit the free-space link by operating an Einstein synchronization link to the ACES module at the ISS. Although we wish to do a frequency comparison between the clock transition interrogation lasers, we need to do time transfer because of the ranging component in the link. This is a requirement of the Einstein Synchronization. Otherwise, we lose the information about the distance between ground station and satellite clock. This automatically means that we have to integrate the tick rate of the clock continuously and repeat the measurements frequently over a number of overhead passes of the ISS. Now what is the tick-rate difference of the two optical clocks that we have to expect? The ratio of the frequency f_{hi} of the clock in the higher location and the frequency f_{lo} at the lower location is

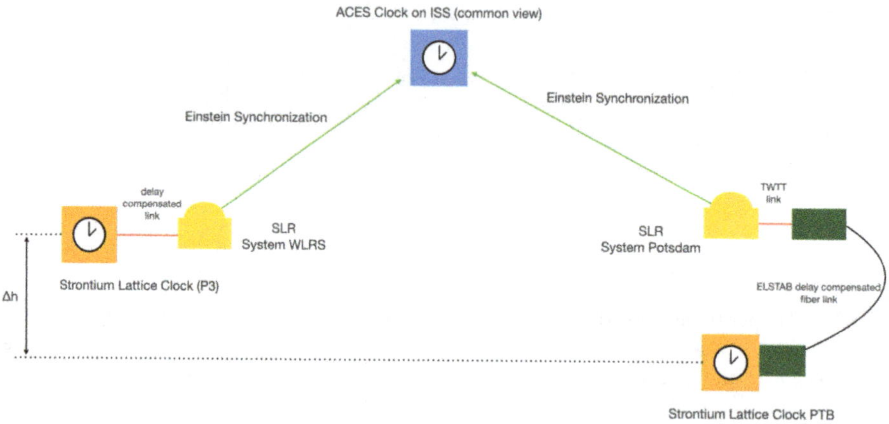

Figure 5.29. Experimental sketch for chronometric leveling utilizing an optical free-space link via satellite.

$$\frac{f_{\text{hi}}}{f_{\text{lo}}} = 1 + \frac{gh}{c^2}. \tag{5.11}$$

With the gravitational acceleration $g = 9.81$ m s^{-2} and the expected height difference of 530 m, we get for the fractional frequency difference $\delta f = 5.785 \times 10^{-14}$. Multiplied with the transition of a strontium lattice clock (429 THz) we obtain a difference in the clock transition of 24.8 Hz. By taking the fractional frequency difference and a division by the number of seconds in a day, we finally obtain the tick-rate difference of 5 ns in a day. Integrating over several days will allow to establish the actual height difference with good resolution. To achieve that, both clocks must run continuously. It remains to be seen over the next 2 years how well the height difference can be established by this concept, utilizing a highly stable clock orbiting in space. In a ground-based local experiment at the top and the bottom of the Tokyo tower [18], a remaining error for the chronometric leveling of $\Delta h = \pm 3.9$ cm has been demonstrated. This compares to a remaining leveling error by GNSS evaluation and laser ranging, carried out on the same platform of $\Delta h = \pm 1.3$ cm.

References

[1] Abshire J B, Sun X, Neumann G, McGarry J, Zagwodzki T, Jester P, Riris H, Zuber M and Smith D 2006 Laser pulses from Earth detected at Mars *2006 Conf. on Lasers and Electro-Optics and 2006 Quantum Electronics and Laser Science Conf.* pp 1–2

[2] Cacciapuoti L *et al* 2024 Atomic clock ensemble in space *J. Phys Conf. Ser.* **2889** 012005

[3] Degnan J J 2002 Asynchronous laser transponders for precise interplanetary ranging and time transfer *J. Geod.* **34** 551–94

[4] Degnan J J 2006 Simulating interplanetary transponder and laser communications experiments via dual station ranging to SLR satellites *Proc. of the 15th Int. Workshop on Laser Ranging* **34** 551–94

[5] Guillemot P, Gasc K, Petitbon I, Samain E, Vrancken P, Weick J, Albanese D, Para F and Torre J M 2006 Time transfer by laser link: The T2L2 experiment on Jason 2 *Proc. of the 2006 IEEE Int. Frequency Control Symp.* pp 5–7

[6] Kodet J, Panek P and Prochazka I 2015 Two-way time transfer via optical fiber providing subpicosecond precision and high temperature stability *Metrologia* **53** 18–26

[7] Kodet J, Schreiber K U, Eckl J, Plötz C, Mähler S, Schüler T, Klügel T and Riepl S 2018 Co-location of space geodetic techniques carried out at the Geodetic Observatory Wettzell using a closure in time and a multi-technique reference target *J. Geod.* **121** 6109–18

[8] Kodet J, Prochazka I, Blazej J, Panek P, Schreiber U and Lauber P 2012 Two way time transfer with picoseconds precision and accuracy *2012 European Frequency and Time Forum* pp 483–7

[9] Levine J 2008 A review of time and frequency transfer methods *Metrologia* **45** S162–74

[10] Liu T, Eckl J J, Steindorfer M, Wang P and Schreiber K U 2021 Accurate ground-to-ground laser time transfer by diffuse reflections from tumbling space debris objects *Metrologia* **58** 025009

[11] Samain E, Exertier P, Guillemot P, Pierron F, Albanese D, Paris J, Torre J M, Petitbon I and Leon S 2009 Time Transfer by Laser Link T2L2 First Results *2009 IEEE Int. Frequency Control Symp. Joint with the 22nd European Frequency and Time forum* pp 194–8

[12] Samain E, Vrancken P, Guillemot P, Fridelance P and Exertier P 2014 Time transfer by laser link (T2L2): characterization and calibration of the flight instrument *Metrologia* **51** 503–15

[13] Schreiber K U and Kodet J 2017 The application of coherent local time for optical time transfer and the quantification of systematic errors in satellite laser ranging *Space Sci. Rev.* **214** 1371

[14] Schreiber K U, Lauber P, Hiener M, Holzapfel B, Michaelis H, Brandl N, Haufe K -H and Neidhardt A 2009 Altimetry and transponder ground simulation experiment *Planet. Space Sci.* **57** 1485–90

[15] Schreiber K U, Prochazka I, Lauber P, Hugentobler U, Schäfer W, Cacciapuoti L and Nasca R 2010 Ground-based demonstration of the European Laser Timing (ELT) experiment *IEEE Trans. Ultrason. Ferroelectr. Freq. Control* **57** 728–37

[16] Smith D E 2006 Two-way laser link over interplanetary distance *Science* **311** 53

[17] Smith D E *et al* 2017 Summary of the results from the lunar orbiter laser altimeter after seven years in lunar orbit *Icarus* **283** 70–91

[18] Takamoto M, Ushijima I, Ohmae N, Yahagi T, Kokado K, Shinkai H and Katori H 2020 Test of general relativity by a pair of transportable optical lattice clocks *Nat. Photon.* **6** 1–7

[19] Vrancken P 2008 Characterization of T2L2 on Jason 2 ocean altimetry satellite and micrometric laser ranging *PhD Thesis* University of Nice-Sophia Antipolis

[20] Zuber M T *et al* 2010 The Lunar Reconnaissance Orbiter Laser Ranging Investigation *Space Sci. Rev.* **150** 63–80

IOP Publishing

The Measurement Techniques of Space Geodesy
Optical technologies

Karl Ulrich Schreiber

Chapter 6

Sagnac interferometry

The rotation rate of the Earth is perceived as very constant. Therefore, it took a long time until our ancestors realized that they were living on a revolving sphere. Furthermore, the domain of our celestial neighbors was considered as the realm of the gods and so the heavens were thought to be a perfect clockwork. The shape of the celestial bodies was assumed to be a perfectly round sphere. Their orbit, once it was recognized, was initially considered to be exactly circular and so on, simply because gods are perfect. Only very reluctantly people began to accept that things were not perfect. Neither had the Earth a perfectly spherical shape, nor was its axis of rotation perpendicular to the ecliptic and the circular orbit turned out to be elliptical. Nothing was easy and the closer we look the more we find irregularities in the motion of the Earth and the cosmos around us. Based on careful observations, a lot of models have been developed over the centuries, so that many interacting mechanisms have been identified and understood. Nevertheless, some interactions between physical causes and the motion of the Earth are not predictable. Their interaction is too complex, or the driving interdependencies are still not sufficiently identified in some parts. The consequence is that we still rely on accurate and highly resolved frequent observations to precisely model the motion and the instantaneous orientation of the Earth in space in order to support navigation on the Earth and in the Solar System and for the development of a better understanding of the complex interactions in the system Earth.

6.1 The Sagnac effect

In principle, there are two ways we can establish the orientation of the Earth and the direction into which the Greenwich meridian is pointing at any given time. We can either look outside and use celestial objects as beacons for the highly resolved measurement of directions, or use a self-contained inertial measurement technique [4]. The advantage of the former method is the long-term stability of its reference markers, namely highly energetic quasars, placed near the edge of our observable

doi:10.1088/978-0-7503-2032-0ch6

Universe. The distance to these objects is so vast that they do not move perceptively in our lifetime. Several hundreds of reasonably 'well behaved' quasars are known and mapped out. They make up a nice stable celestial reference frame, against which we can observe the variability of the Earth motion and orientation. The alternative method is to employ an inertial measurement technique, such as a gyroscope [25]. If a suitable technique could be harnessed, the measurements would be relative to the local Universe and entirely independent on anything else around us and they would be permanent. While the direction measurement technique fulfills the important requirement of long-term stability, the gyroscope method carries the promise of permanent availability and high temporal resolution, in particular when it is based on optical waves rather than a mechanical spinning mass.

To understand how optical waves can be used for the determination of angular velocities (or angles), we do a thought experiment as sketched out in figure 6.1. We take a coherent light source (i.e. a laser) from which we obtain a single-frequency beam with the wavelength λ. When the beam hits a perfect circular apparatus, it separates into two equal beams, one going around the perimeter of the circle in the clockwise sense, the other in the counterclockwise sense. At the opposite side, where the beams meet again, they are superimposed and projected onto a suitable photodetector. When we compare the exit beam to the input beam, we observe no change, whatsoever. The beam path for both beams has been identical in length, therefore the phase difference is $\Delta\varphi = 0$. The situation changes if we rotate the circular apparatus while the beams are progressing around the perimeter in both directions. This is illustrated in figure 6.2.

When the instrument is rotating in the counterclockwise direction around an axis, which is perpendicular on the sketch, we see that while the laser beams are traveling around the perimeter, the clockwise beam has a shorter path. It sees the point of recombination moving toward it. The counterclockwise beam sees the point of recombination moving away from it. The result is a phase shift for the recombined beam. We note that this is a consequence of the finite and constant vacuum velocity

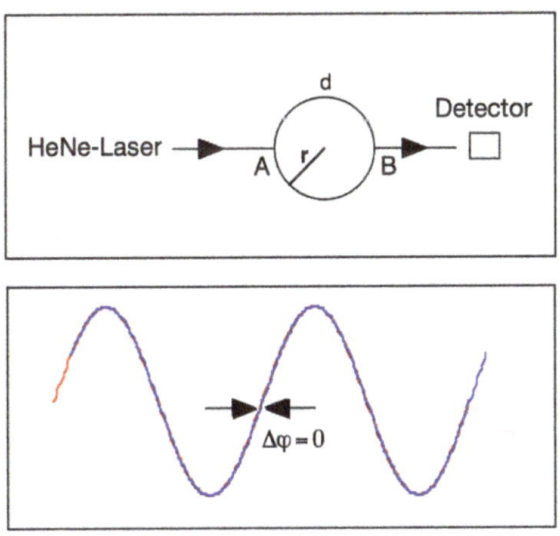

Figure 6.1. Illustration of a thought experiment for a simple demonstration of the Sagnac effect: I. The nonrotating instrument.

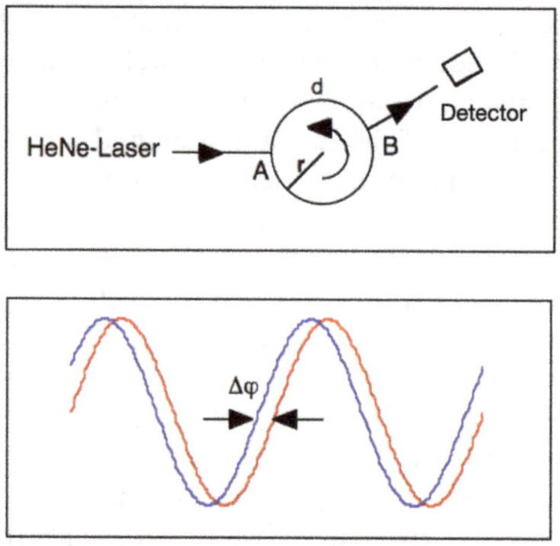

Figure 6.2. Illustration of a thought experiment for a simple demonstration of the Sagnac effect: II. The rotating instrument.

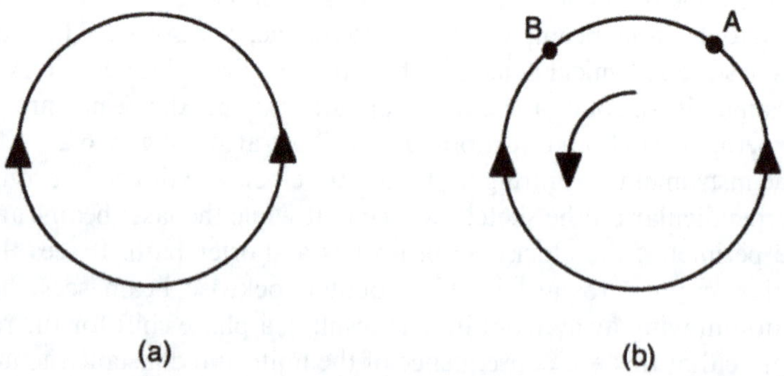

Figure 6.3. A simplified depiction for the determination of the Sagnac equation. The interferometer at rest is shown as (a), while the rotating interferometer is depicted as (b).

of light, which makes this measurement concept an ideal candidate for precise and accurate determinations of rotations. Furthermore, the actual value of the phase shift $\Delta\varphi$ is strictly linearly proportional to the rate of rotation Ω. Another interesting aspect of our thought experiment is the fact that we have considered the light beams as constant and the apparatus as moving along with the rotating platform, or in other words, we have considered the light pattern as fixed with respect to the Universe and the apparatus as strapped down to the object of interest, whose spinning motion we wish to quantify. This is the typical way of operating such a Sagnac gyroscope. How large is the Sagnac effect? To explore this, we simplify our thought experiment to the sketch shown in figure 6.3.

For the apparatus at rest, as illustrated in figure 6.3(a), we can calculate the time t the light takes to go around the circle as

$$t = \frac{2\pi r}{c}. \tag{6.1}$$

When we rotate the apparatus around a vector vertical on the circular contour, we estimate for the length of the beam path traveled to complete a whole circle

$$ct_+ = 2\pi r + (\Omega r)t_+. \tag{6.2}$$

The origin of the beam path is point A that has moved over to the position B in figure 6.3(b) while the beam is going around the circle. The entire path traveled by the beam is therefore a full circle plus the little section between the points A and B. Converted to the travel time t_+ of the corotating beam, we obtain

$$t_+ = \frac{2\pi r}{c - \Omega r}. \tag{6.3}$$

We can do the same considerations for the antirotating beam, which is going from point A to B in figure 6.3 without completing a full circle.

$$ct_- = 2\pi r - (\Omega r)t_- \quad \text{and} \quad t_- = \frac{2\pi r}{c + \Omega r}. \tag{6.4}$$

An outside observer notices the path difference ΔL between the two beams as

$$\Delta L = c(t_+ - t_-) = 2\pi rc\left(\frac{1}{c - \Omega r} - \frac{1}{c + \Omega r}\right). \tag{6.5}$$

By applying the rules of the binomial formulas, this can be simplified to

$$\Delta L = 2\pi rc\frac{2\Omega r}{c^2 - \Omega^2 r^2}, \tag{6.6}$$

We use the approximation $\Omega r \ll c$ to simplify the expression for ΔL further

$$\Delta L \approx \frac{4\pi\Omega r^2}{c}. \tag{6.7}$$

To get to the time difference between the two beams reaching the position B we write

$$\delta t = \frac{\Delta L}{c} = \frac{4\pi\Omega r^2}{c^2} = \frac{4A}{c^2}\mathbf{n} \cdot \boldsymbol{\Omega}. \tag{6.8}$$

Coherent light has the shape of a periodic sinusoidal signal, hence we can express this difference as a phase difference

$$\delta\phi = \frac{2\pi}{\lambda}\Delta L = \frac{8\pi A\Omega}{\lambda c} = \frac{8\pi A}{\lambda c}\mathbf{n} \cdot \boldsymbol{\Omega}. \tag{6.9}$$

The result is the *passive* Sagnac equation. Passive in this context means that the light source is outside the interferometer. Both beams have the same frequency and are traveling in opposite directions. The observable is a phase shift, which depends on the rate of rotation as well as the scale factor. The combination of enclosed area, the corresponding perimeter and the wavelength are making the scale factor up. A larger scale factor provides higher sensitivity, but becomes progressively more difficult to maintain constant.

In contrast to the passive Sagnac equation, we also have the *active* Sagnac equation, which has the laser frequency generation inbuilt as part of the ring cavity. The operating frequency is determined by the length of the cavity. We find that the co- and antirotating beam each experience the identical delay as in the passive system. In order to obtain coherent laser amplification the system has to satisfy the lasing condition, which requires that there is an integer number of waves around the cavity ($L = n\lambda$). A consequence of the lasing condition is

$$\frac{\Delta\nu}{\nu} = \frac{\Delta L}{L}. \tag{6.10}$$

We reorganize the equation and use the relation $\nu = c/\lambda$ to obtain

$$\Delta\nu = \frac{\Delta L}{L}\nu = \frac{c}{\lambda L}\Delta L = \frac{c}{\lambda L}\left(\frac{\lambda}{2\pi}\delta\phi\right). \tag{6.11}$$

Taking equation (6.9) to substitute $\delta\phi$ we finally get the Sagnac equation for the active case

$$\Delta\nu = \frac{c}{2\pi L}\left(\frac{8\pi A\Omega}{\lambda c}\right) = \frac{4A}{\lambda L}\mathbf{n} \cdot \boldsymbol{\Omega}. \tag{6.12}$$

We note that the phase difference $\Delta\varphi$ has morphed into a frequency difference Δf, which is technically much easier to evaluate accurately. The expression $4A/(\lambda P)$ is the scale factor again. It depends only on the size of the gyroscope and defines the ultimately achievable resolution. It is important to note that square gyros smaller than 16 m^2 are too small for high-resolution earth rotation sensing. Instruments much larger than that are usually not stable enough. We remind ourselves that we are after the rotation rate of the Earth (Ω_E) at a resolution of $\Delta\Omega/\Omega_E = 10^{-9}$. That means the scale factor needs to be even more stable than that to ensure that we measure a geophysical signal and not an instrumental stability limit.

6.2 A brief history of Sagnac interferometers

Historically, Sagnac interferometry reaches back to the days of the search for the ether at the end of the 19th century. Analogous to the propagation of sound in the atmosphere, light and later radio waves were thought to be traveling in a yet-unknown medium. Furthermore, in these days before the theory of special relativity, time and space were still considered absolute. The rotation rate of the Earth around its spin axis and the orbital angular momentum of the Earth around the Sun should therefore provide a measurable length difference in an optical interferometer with two arms pointing in orthogonal directions, namely east–west and north–south. This was the basic idea behind the most famous 'zero effect' experiment in physics, the Michelson–Morley experiment [19]. The published result from their original paper is shown in figure 6.4. The dashed line represents the expected measurement result, while the solid line provides the measurement. Since the distance between Earth and

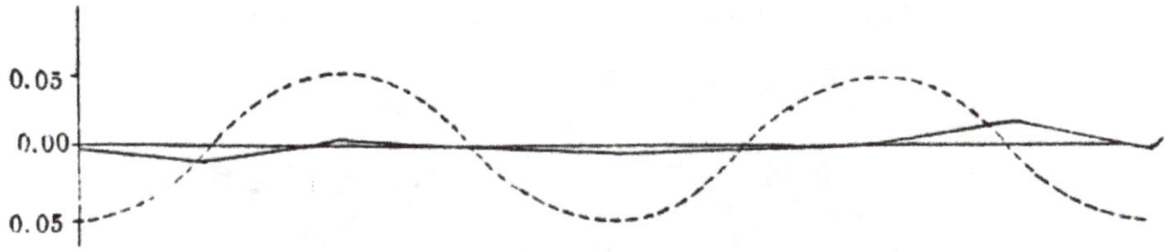

Figure 6.4. The result of the Michelson–Morley experiment. The dashed line marks the expected result and the solid line represents the measurements. It is important to note that the dashed line assumes an orbital velocity of the Earth of 8 km s^{-1}, the best value known at the time.

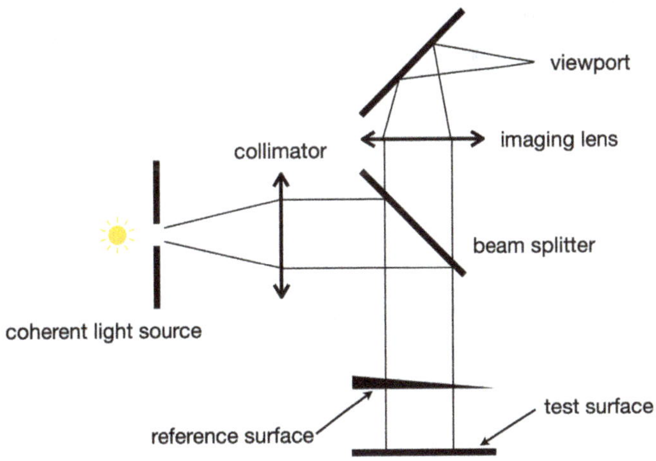

Figure 6.5. Illustration of the construction of a Fizeau interferometer. Today one would use laser light as a source, which of course was not available at the time of the experiment.

Sun was not well known at the time of the experiment, the orbital velocity was assumed to be 8 km s^{-1}, while the correct value amounts to 30 km s^{-1}. Today we would certainly also include the orbital angular momentum of the Sun around the galactic center, which was not known at the time of this experiment either and results in an orbital velocity of the Earth of 220 km s^{-1}. The only remaining conclusion of this zero result was the idea that the ether might be dragged along with the Earth.

Sir Oliver Lodge pursued this idea by devising an experiment in 1893 involving a whirling machine to prove the entrainment of the ether [12]. He remarked:

… *There was a controversy at the time as to whether the velocity of light as determined on the Earth surface was in any way affected by the motion of the Earth; or since no one doubted that the ether was the vehicle conveying light, it might be expressed as whether the ether was carried along by the Earth at all* …

His apparatus contained a Fizeau interferometer in a very rugged construction, which had fast rotating plates of 1 m in diameter mounted at very short distance above and below the interferometer. A Fizeau interferometer uses a coherent light source and two surfaces at relative close distance, whose reflections are interfered, so that fringes can be studied at the viewport. Figure 6.5 illustrates the setup.

Figure 6.6. The whirling machine of Sir Oliver Lodge (left), built in search of the ether and entrained by two rapidly moving plates above and below a Fizeau interferometer. (Source: Cadbury Research Library OJL5/2, Special Collections, University of Birmingham.)

A suitable light source such as a mercury arc lamp and a pinhole generates a sufficiently coherent monochromatic light beam, which is collimated and reflected by a beam splitter toward the surface of interest and passing through the partially transmitting reference surface on the way. Once the reflected light with the interference pattern of the two surfaces has passed through the beam splitter, it is imaged onto the viewport. If the ether can be dragged along by moving masses, it would change the observed fringe pattern. Sir Oliver Lodge achieved rotation rates of up to 3000 revolutions per minute. Apart from the experimental problem of keeping the vibrations of the revolving plates away from his optical interferometer, which was no mean feat, he saw nothing. Figure 6.6 provides a contemporary depiction of the whirling machine.

Disappointingly for him, the experiment did not show any effect of the ether either, which put the ether hypothesis into a precarious position. So the question arose concering whether the resolution of the experiment was sufficient. The next notable experiment was carried out by George Sagnac in 1913. His intention was to refute special relativity, which had emerged in the mean time and made the ether theory obsolete. This experiment differed from the earlier attempts by rotating the interferometer itself, an undertaking that is experimentally extremely awkward. The interferometer must be shielded carefully from the effects of centrifugal forces, which may bend some of the mechanics and thus introduce some effect proportional to the angular velocity. His device employed a completely reciprocal beam path around a contour enclosing an area, similar to our thought experiment at the beginning of this section. Figure 6.7 shows a sketch of his famous experiment [22]. The fact that he chose this arrangement helped a lot in avoiding systematic errors from interferometer deformation. It would be reciprocal for both beams.

The instrument featured a coherent light source, which is shown on the lower-left side. The beam is collimated and split into two beams at a beam splitter. Each of the

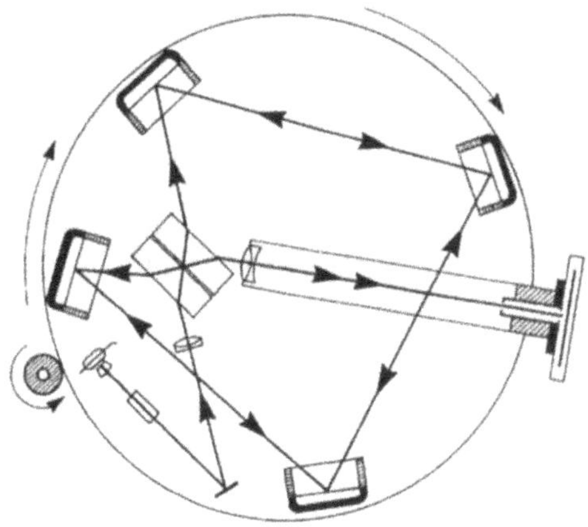

Figure 6.7. The layout of the classical Sagnac experiment. The two beams travel on the identical path, but in different directions around a contour enclosing an area. (Figure adapted from [22].)

beams travels around the contour set up by the four mirrors, before they are recombined at the beam splitter and imaged on a photo plate. The area enclosed by the beams amounted to $A = 0.086\ \text{m}^2$ and the experiment was rotated at two revolutions per second. George Sagnac observed an angular velocity related shift of the fringes by 0.07 ± 0.01 fringes.

$$\delta\phi = \frac{8\pi A}{\lambda c}\boldsymbol{n} \cdot \boldsymbol{\Omega}. \tag{6.13}$$

The observed result is compatible with equation (6.13). It requires, however, that there is no ether entrainment involved. In fact, it also requires that the velocity of light is finite and constant, something that George Sagnac completely refused throughout his life. Although intended to refute the theory of special relativity, the experiment turned out to provide overwhelming proof. Taken together, the Michelson–Morley and Sagnac experiments did not leave any room for the existence of an ether. The famous Maxwell equations of modern electrodynamics render the existence of an ether as entirely unnecessary. This historical development is a nice example how science develops, sharpens and progresses over the years. Apart from smart ideas and experiments, there is a lot of trial and error involved.

There is still another important milestone toward modern highly sensitive optical rotation sensing devices and this is the Michelson–Gale Experiment, published in 1925 [17, 18]. It was still attached to the search for the ether, but it was eventually not necessary to disprove the ether theory. All the necessary discoveries to refute this theory had been available before this experiment, but it still took time for the consequences to be widely accepted in the scientific community. Even Albert Michelson was a bit skeptical about this experiment before it was completed. He is quoted to have said:

Figure 6.8. The Michelson–Gale Experiment in Clearing (Illinois) in 1925. Although motivated by the search for the ether, it was a groundbreaking experiment because it marked the transition from a tabletop experiment to a large-scale application, something that is characteristic for space geodesy.

Well, gentlemen, we will undertake this, although my conviction is strong that we shall prove only that the Earth rotates on its axis, a conclusion which I think we may be said to be sure of already.

We are looking at this experiment with admiration for a very different reason. The apparatus was made highly sensitive by scaling the area up or in a better formulation, which resulted in the scale factor of the instrument being vastly increased. It allowed the rotation rate of the Earth to be obtained and yet it was strapped down to the ground. There was no turntable involved. Figure 6.8 shows the outline of the experiment from the 1925 publication [18]. The impressive size of the rectangular optical beam arrangement is reminiscent of the gravitational wave detection experiments about one hundred years later. The interferometer featured 603 m of length in the east to west direction and 334 m in the north to south direction. The carefully surveyed mirror positions were enclosed by sealed-off water pipes with a diameter of 30.5 cm. This was necessary in order to pump the air out of the system, which otherwise would disturb the interference fringes by turbulence. A stable interference pattern was achieved for pressures below 34 hPa. A delicate lever system allowed the adjustment of the mirrors in the corner boxes under vacuum. The main interferometer was formed by the mirrors A, D, E and F. The mirrors A, B and C acted as beam splitters, and all other mirrors were highly reflective. The light came from an arc lamp behind a slit. It was projected into the coarsely evacuated interferometer with a collimating lens. The fringe shift was established to $\delta\phi = 0.23 \pm 0.005$. This corresponds to the estimation of the rotation rate of the Earth with an error of 2%, an excellent achievement for the early 20th century. The alignment of the mirrors required a stable construction for the pads where the corner boxes were placed on. Furthermore, the team had to align the optical path to within a few seconds of arc over a large distance of several hundred meters. A serious problem remained, however. While George Sagnac could either ramp up or slow down the angular velocity of his turntable, Albert Michelson could not alter the rotation rate of the Earth. So how could he prove that the fringe pattern he saw was indeed coming from the Earth's rotation. This is where the beam path between the mirrors A, B, C and D became relevant and served as a calibration. Since the enclosed area of this part of the interferometer was too small to matter, it did not show any rotation related fringe shift but used the same optical alignment for the critical elements in this setup. The measurements therefore compared a large area interferometer to a zero area interferometer, all in one optical setup. Upscaling turned out to be the most effective way to increase the sensitivity of a Sagnac type of interferometer and this has not changed.

Space geodesy today does not require a 2% resolution of the rotation rate of the Earth. The rotation rate of the Earth acts as the linking element between the celestial frame of reference and the terrestrial frame of reference, between the celestial (ICRF) and terrestrial (ITRF) reference frames. We require positions and bearing on the surface of the Earth, but the objects of observation are either astronomical objects or satellites in an earth orbit, all of which do not have any link to the spin of the Earth. For that matter we require seven orders of magnitude more sensitivity than the Michelson–Gale interferometer can deliver. Furthermore, we also require an enormous instrumental stability because a large number of geophysical processes act over timeframes of many months. With the advent of lasers as monochromatic, highly coherent light sources and the invention of ion beam sputtering as a new optical coating process for the manufacturing of mirrors with extremely low losses, the necessary improvements became available. The optical maser, today known as the laser (**L**ight **A**mplification by **S**timulated **E**mission **R**adiation), was already predicted by quantum mechanics in the early stages of the 20th century. The maser, utilizing the same physical properties for microwaves, was realized in the forties of the 20th century. In 1960, optical pumping had been successfully demonstrated on a three-level ruby system by Theodore Maiman [14]. The suitability of this light source for rotation sensing was quickly recognized. Three years later W Macek and D Davies, who worked for the Sperry Gyroscope Company Division, published the first result of an active helium neon Sagnac interferometer [13]. The losses in this early ring laser cavity were prohibitively high, which caused the laser frequencies in the cavity to degenerate. This phenomenon of a zero beat in the presence of rotation became known as the lock-in effect [21]. In order for rotation sensing to occur, the system had to be put on a turntable. Nevertheless, this proof of concept had significant advantages. There were no moving mechanical parts in it and the system covered an enormous dynamical range. So it was no surprise that a large commercial multimillion dollar effort was prompted in the seventies to develop a moderate sized instrument, suitable for aircraft dynamical attitude control. From the eighties onward, a ring laser gyro inertial measurement unit (IMU) could be found in almost any new aircraft and particular stable high-end units in submarines. An IMU comprises three laser gyros for pitch, yaw and roll and three accelerometers in an orthogonal configuration. Since upscaling promises higher resolution, this commercial and dual-use development had a spin-off in search for a superstable and ultrasensitive gyroscope for geophysical applications [4]. This effort is covered in more detail in [25, 30].

6.3 Types of Sagnac interferometers

We saw in the beginning of this chapter that the Sagnac effect can be exploited by utilizing an external laser source. We call this type of interferometer passive because it operates on a 'ready made' light source and the mirror assembly only provides a beam path. After the recombination of the two counterpropagating beams we get the fringe pattern, which contains the information about the rotation rate. Alternatively, the laser gain medium can be part of the interferometer itself, so that the beam path becomes a laser cavity. Then the oscillation frequency in the two different directions differs slightly and the beat note between the two is proportional to the rotation rate. When the laser is external to the interferometer, we have the advantage of an empty cavity. There is nothing in the way that can cause a nonreciprocal cavity contribution. The precision or the detected fringe shift depends very much on the contrast of the interferogram and that again depends on the beam power in the instrument, which in principle can be made very large.

The active system on the contrary has the advantage that the linewidth of the excited laser beam can become extremely narrow in a ring cavity with a high-quality factor, Q. We can write $Q = \omega\tau$, where ω is the optical frequency and τ is the cavity ring-down time[1], which is the time that the light takes to vanish from the cavity when the light source is turned off. Typical values for modern low-loss mirrors are in the range of 1 ms for a cavity of 16 m length. A narrow linewidth automatically means that the beat note is sharply defined and the precision of the detected beat note of the active gyroscope is very high [30]

$$\Delta(\delta f) = \sqrt{\frac{hf_0^3}{Q^2 P_0 T}}. \tag{6.14}$$

Based on that we can find the sensitivity limit of the laser gyroscope for rotation rates as

$$\delta\Omega = \frac{cP}{4AQ}\sqrt{\frac{hf_0}{P_0 T}}. \tag{6.15}$$

This resolution limit is also known as the **quantum limit** of a laser gyroscope. P is the perimeter, c the velocity of light, A the area enclosed by the laser beams, Q is defined as above, h the Planck constant, f_0 the optical frequency of the laser beams, P_0 the light power on the detector and finally T the time of integration. We note that the quantum limit is not a fixed number, it depends on the setup and the operation characteristics. However, equation (6.15) emphasizes the importance of a high Q very clearly. Figure 6.9 provides the sensor resolution of the large ring laser G at the Geodetic Observatory Wettzell. It is currently the most sensitive and stable earth-observing gyroscope. After 4 h of integration, it reaches a sensitivity level of 4×10^{-13} rad s^{-1}. Ring lasers in aircrafts are orders of magnitude less sensitive, but feature fast update

[1] The ring-down time is the time that the intensity of the laser takes to drop to a value of $1/e$ when the gain is suddenly switched off.

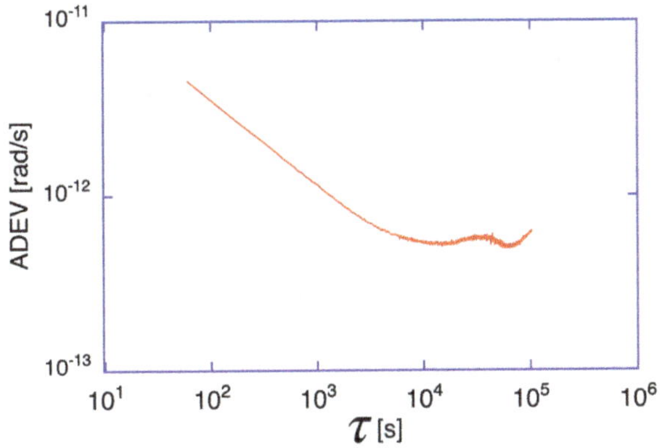

Figure 6.9. The Allan deviation of the G ring laser at the Geodetic Observatory Wettzell. After 4 h of integration, the system levels off in a flicker floor. For longer integration times we would see a slope increasing as $\sqrt{t}$, indicating a random walk, as the cavity drift takes over.

rates in the range of kHz and can handle angular changes of more than 100° s^{-1}. They are made for a different purpose, namely the real-time attitude control of aircrafts, while our geodetic instruments are designed for ultimate resolution and high stability. While navigation gyros are strapped down to the aircraft in order to sense its motion reliably, geodetic gyroscopes are strapped down to the Earth in order to capture small irregularities in the rotational motion. One would hope that the Allan deviation of a ring laser gyroscope would average down almost indefinitely, but this is not observed. The G ring laser, probably the currently most stable ring laser gyroscope, reaches a resolution minimum after 4 h. This is caused by tiny asymmetries in the counter-propagating laser beams in the cavity and not yet fully understood.

The linewidth of the passive system Δ_{ν_c} is of the order of 200 Hz, while the active system features values around $\Delta\nu_L \approx 7$ μHz and this can be computed by

$$\Delta_{\nu_c} = \Delta_{\text{FSR}}\frac{1-R}{\pi\sqrt{R}} \quad \text{with} \quad \Delta_{\text{FSR}} = \frac{c}{L}, \tag{6.16}$$

where Δ_{FSR} is the free spectral range, R the mirror reflectivity, c the velocity of light and L the length of the resonator [16].

$$\Delta\nu_L = \frac{N_2}{N_2 - N_1}\frac{2\pi h f_0 \Delta\nu_c^2}{P_L} \tag{6.17}$$

$\Delta\nu_L$ is also known as the Schawlow–Townes linewidth limit with $N_{1,2}$ the population numbers of the lower and the upper laser state, h the Planck constant, f_0 the laser frequency, P_L the laser power and the cavity resonance width $\Delta\nu_c$ defined as before. This extremely narrow linewidth is by far the greatest advantage of the HeNe neutral gas laser. It provides an incredible resolution for an active ring laser. Nevertheless, this has also a downside. As the main disadvantage of the active system we have to accept the gain medium in the cavity as part of the sensor. That means, the cavity is not empty. Therefore, we need to deal with nonlinear plasma dynamics [31] and it becomes more difficult to obtain the correct scale factor. Since we require a resolution in the range of 1 ppb, we face a challenge here. The gain medium can also exacerbate systematic biases from coupling of the two laser beams and other asymmetries in the cavity behavior. Usually, one would expect that the two counterpropagating laser beams excited in the active cavity by sharing the same gain medium will exhibit the same intensity. However, this is not the case. In all the gyroscopes that we have built, we always had an asymmetry in the beam intensities at levels between 1% and 5%. That means, either the clockwise cavity or the counterclockwise cavity experienced more losses than the other and errors common to both beams do not entirely cancel out. We can distinguish three different approaches to the realization of a Sagnac interferometer, keeping in mind that the ability of rotation sensing is entirely a property of the beam guiding contour, enclosing an arbitrarily shaped area [25].

Figure 6.10 illustrates the three existing types of optical Sagnac interferometers symbolically. Only one of them (A) marks the classical realization of a phase shift measurement device. An external laser source hits a 1:1 beam splitter, so that two beams go around the contour before they recombine at the beam splitter and generate the interference fringes, which then can be evaluated. The other two systems exploit a resonant cavity. The sketch in the middle illustrate an external laser that is injected into the cavity to excite and lock to a cavity resonance (B). In order to keep the illustration simple, only one of the two necessary locking loops is indicated. The third sketch depicts an active ring laser cavity (C). Here the gain medium is part of the cavity. It is important that the two excited laser beams in this gain section are independent of each other and as much decoupled as possible. In terms of required technology, the active laser gyro is the simplest of all concepts. While the cavity provides the rotation sensing ability of these instruments, the

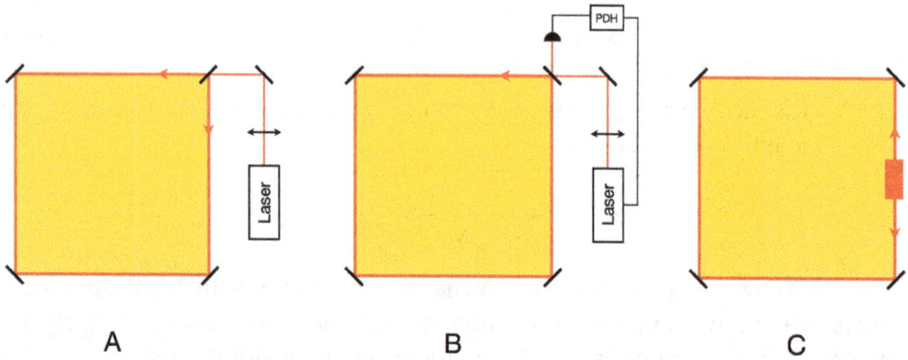

Figure 6.10. A passive Sagnac interferometer has one light source and a beam splitter divides the beams into a clockwise and anticlockwise beam (A). The result is a phase shift, proportional to the experienced rate of rotation. The external laser can also be locked to the resonance of the empty cavity (B). This requires a laser lock by a feedback system for both directions. A ring laser gyroscope finally has the gain medium inside the cavity (C).

applications of the laser technology is only required to extract the actual rotation motion, which the whole cavity is subjected to. At the time of writing in 2025, the laser gyro provides the most advanced results for rotation sensing in space geodesy.

6.4 The ring laser gyroscope

Once the working principle of a ring laser gyro had been demonstrated by Macek and Davies [13], its potential for aircraft attitude control was very quickly realized. Until then, mechanical gyros had been in use, which were highly complex, expensive and vulnerable instruments. During the seventies there was a large effort and many million US dollars spent on the development of a ring laser-based IMU. Such a system contains three gyroscopes and three accelerometers in an orthogonal configuration. Once set to an initial position, such an IMU measures the rotation rates and integrates the angles through which the aircraft is moving. At the same time it integrates the measured accelerations twice to get the distance and combines the measurements of all six sensors to the instantaneous position of the aircraft in space and the path it has taken. While the system certainly allows to navigate, the main objective is to provide the attitude control. Therefore, such a system is indispensable for autopilot operations. The necessary fundamental theory for the ring laser physics dates back to this era [2, 32] and is still fully applicable today.

Ring lasers for the purpose of space geodesy emerged in the early nineties [3, 4]. This interest arose from the desire to have a continuous data set for the rotation rate of the Earth and the orientation of the Earth in space. This knowledge is required to compute the terrestrial reference frame for the Earth. Starting from a stable and fixed celestial reference frame, it is possible to perform the transformation to the rotating surface of the Earth. Space geodesy uses quasar positions as stable cosmic markers and observes orbiting satellites in order to measure positions and distances on Earth. Gravitational attraction from Sun and Moon give rise to polar motion. Free oscillations and deformation processes on the Earth are complicating this transformation. Furthermore, none of these observations are continuous. From this point of view it is very desirable to develop an observation technology, which could fill this gap [4]. From various candidates, the ring laser gyro is the best performing approach today, but still full of challenges. In chapter 2 we saw that a maser makes an excellent clock in the short term, but drifts in the long term, so that the interrogation of an atomic state ^{133}Cs is required to provide the long-term stability. A laser is an optical maser and has the same general properties, albeit at vastly higher frequencies. The excellent short-term stability of a laser gyroscope is apparent from figure 6.9, but until today it cannot be extended far enough to capture seasonal variations in the Earth motion, such as the El Nino Southern Oscillation (ENSO). In the following we take a look at ring laser technology. A detailed report for further reading can be found in [25, 29].

6.4.1 Design of a ring laser

The design of a ring laser gyro for geodetic observations is governed by three considerations:

- Make the **scale factor** as large as practically possible in order to obtain high sensitivity, while still maintaining high stability of the optical resonator.
- Reduce the **losses** in the cavity to the barest attainable minimum. Low resonator losses provide a narrow laser linewidth and high photon flux.
- Place the gyroscope on a very **stable foundation**. We require a representative point of contact from the sensor to the body of the Earth, since we are not interested in local rotation. We need to evaluate a global quantity.

High symmetry of the sensor arrangement is desirable in order to minimize the effect of potential systematic errors. We have two counterpropagating laser beams in the cavity, which we beat against each other. High sensor symmetry supports the effect that systematics are acting on both beams in the same way. Most of that will then cancel out as a common mode bias in the interferogram. The two requirements of a large-scale factor and low losses are best met in a square cavity. A triangle has fewer mirrors and hence three instead of four items that generate losses, but the area is significantly smaller and therefore the scale factor is reduced. Arrangements with more than four mirrors add more losses with each extra mirror, while the improvement of the scale factor is only very marginal.

A typical suitable realization is shown in figure 6.11. It depicts the 1 m^2 C-II ring laser (A) when it was fabricated in 1996 and a sketch of the experimental design, which shows some of the functional components of the general ring laser operation. C-II belongs to the category of monolithic gyroscopes. It is entirely made from Zerodur, a low thermal expansion glass ceramic. The thermal expansion coefficient $\alpha = 2 \times 10^{-9}$ per °C has been made as small as technically possible, so that the scale

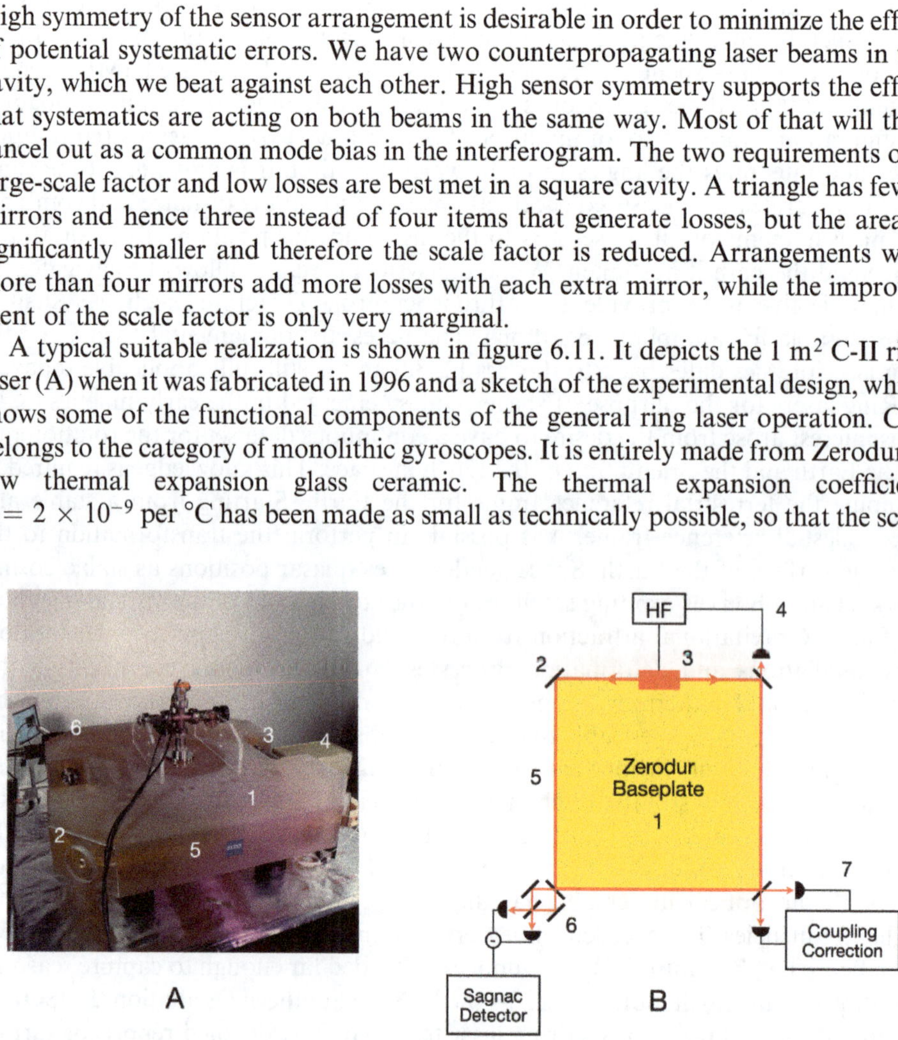

Figure 6.11. The C-II ring laser (A) (Image credit: ZEISS Archive), and the basic ring laser setup (B). A solid Zerodur baseplate provides the body of gyroscope (1). Super low-loss mirrors provide a narrow linewidth (2). A capillary with 4 mm diameter provides the section where laser activity can be exited by a gain controlled radio frequency transmitter (4). The beam path is completely inside a ultrahigh vacuum enclosing (5). The two excited laser beams are recombined and detected (6). In order to correct for frequency coupling, the intensities of each beam have to be recorded (7).

factor does not change with ambient temperature variations. On top of that, C-II had been placed on a solid concrete monument in an underground cavern facility in New Zealand with naturally low-temperature variations. To operate the gyroscope, the beam path inside the Zerodur structure had to be evacuated to values below 10^{-6} hPa before it was filled with 0.2 hPa of to different isotopes of neon, namely ^{20}Ne and ^{22}Ne in a 1:1 mass ratio. The connection interface for the ultrahigh vacuum (UHV) pump is located on top of the Zerodur body of the gyroscope in figure 6.11(A). Diagonal bore holes connect the pump interface internally to the beam path around the perimeter of the sensor block. When the neon is in place, helium is added to obtain a total pressure of the laser gas between 1 and 10 hPa. The highest gain of the helium neon mixture is obtained around 2 hPa. Higher pressures yield lower gain and this is an important property for the ring laser operation. For proper rotation sensing, we require a single longitudinal $TEM_{0,0}$ laser mode to be present in the cavity. The requirement of minimal losses does not allow to have any frequency selection device inside the cavity. A four-meter-long cavity has a longitudinal mode spacing of 75 MHz and the gain curve of the HeNe laser is approximately 2 GHz wide. Without filtering this would result in up to 26 different $TEM_{0,0}$ laser lines operating at the same time and this would make sensitive rotation sensing impossible. Since the gain curve has an envelope composed of a Lorentzian and Gaussian curve, usually referred to as a Voight profile, the laser frequencies close to the line center have higher gain and the ones closer to the ends of the gain curve experience lower gain. The laser gain is excited by the high-frequency generator, radiating 2 watts of say 50 MHz of rf into the capillary, causing a plasma ignition in the process. This provides the necessary inversion in the neon from which the laser radiation is obtained. When we reduce the gain in the capillary by decreasing the injected rf power, the gain curve gets smaller and the number of laser modes reduces. Unfortunately, the plasma extinguishes before single- mode operation is achieved. By overpressuring the laser cavity, we make the laser gain less efficient, so that the desired single-laser mode operation can be achieved with the plasma still on. This process is known as gain starvation. The beam extraction and superposition (6) provide the beat note between the two laser counterpropagating laser modes of the same longitudinal mode index. When the signals are recombined on a 1:1 beam splitter, we get the interferogram twice. Subtracting one from the other doubles the amplitude, as the two interferogram are in antiphase. However, more importantly, common mode noise from the detection assembly is removed from the waveform in this way. The result is a clear and perfect sine waveform, which is exactly proportional to the rate of rotation, experienced by the complete device. We note that there are no moving elements in the apparatus. Rotation sensing is purely achieved by detecting the effective difference in resonator length of the ring cavity between the two senses of propagation. C-II has a perimeter of 4 m and is located at midlatitude on the southern hemisphere. So this difference turns out to be around 1.5 pm, corresponding to a beat note of 79.4 Hz out of a oscillation frequency of 474 THz.

Figure 6.12 presents what the analog to digital (A/D) converter processes. The waveform is detected with a photomultiplier or a photodiode, digitized with 24 bit, time stamped and then averaged over 1 min before it is processed by a high-resolution frequency estimator, based on the Hilbert transform of the digitized

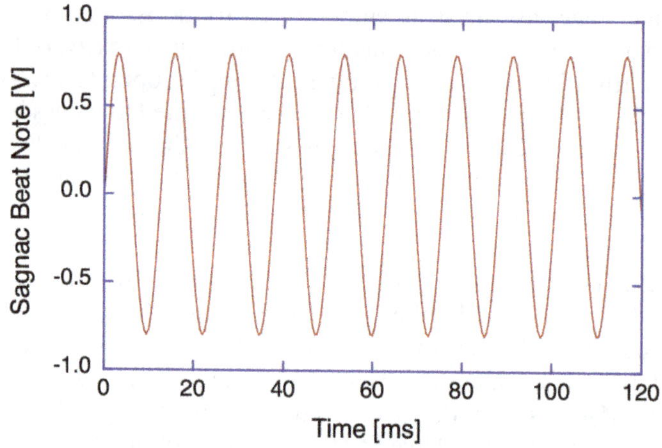

Figure 6.12. Example of what the recorded beat note of the C-II ring laser looks like. The waveform is captured by a 24 bit A/D converter and processed with a numerical frequency estimator.

voltage [25]. Auxiliary signals such as temperature, atmospheric pressure, local tilts and the laser intensities are recorded in parallel. This data is used for the analysis and the correction of the observations.

6.4.2 Error sources in Sagnac interferometers

A single-component gyroscope only provides one component of the three-dimensional Earth rotation vector. We recall equation (6.12), replacing the inner product and also allow for subtle systematics in our measurements

$$\Delta\nu = \frac{4A}{\lambda L}\Omega \sin\theta + f_{\text{nr}}. \tag{6.18}$$

The scale factor is defined as before and $\Omega \sin\theta$ represents the projection of the instantaneous rotation axis of the Earth onto the ring laser normal, with θ the latitude of the sensor location. The value f_{nr} captures nonreciprocal frequency components of the two interfered laser beams. What is that? In the previous paragraphs, we noted that the two oscillating laser frequencies should be absolutely identical apart from the difference induced by the rotating body of the gyroscope. Among other effects, this would require that both beams experience the same gain in the gain medium. Consequently, one would think that the measured intensities are the same. However, this is not the case. We have built more than 10 instruments and none of them exhibits equal intensities. Even the 16 m^2 G ring laser does not run on equal intensities. There is a small difference of less than 1% present. In smaller rings with higher losses, we have seen asymmetries of more than 5%. Furthermore, these differences are not constant. Figure 6.13 shows a measurement series for the losses of the G ring laser. The observed variation is mostly an effect of subtle temperature changes, which stayed well below 0.1° C. If the losses are not the same, we cannot expect that the resultant beat frequency is unperturbed. A tiny shift in the resonant

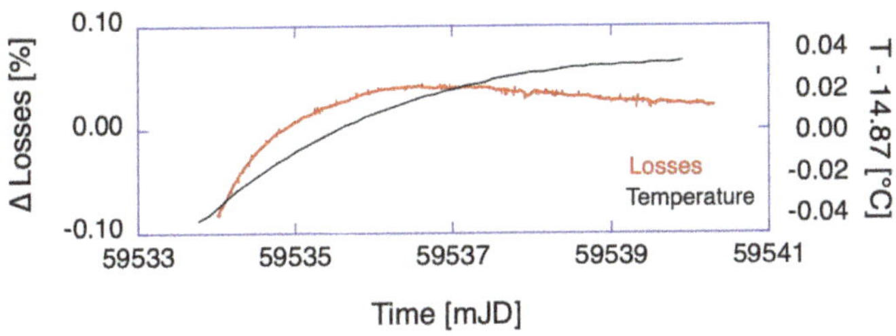

Figure 6.13. The change of the losses of the G ring laser in relation to the simultaneously recorded temperature variations. Although temperature is involved, it is not the only source for the observed drift.

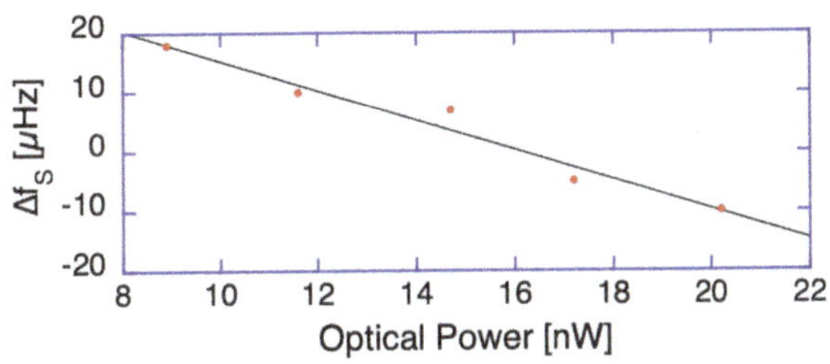

Figure 6.14. The observed shift in the Sagnac frequency as a function of the operation laser power. A value of -2.5 μHZ nW^{-1} has been established.

frequency of either the clockwise oscillator or the counterclockwise oscillator will cause a deviation from the true beat note incurred by the experienced rotation rate. If we consider a tiny shift of 0.5 mHz as an example, this would cause an offset for the Earth rotation rate of 0.14%. Relative to the oscillating optical frequency that would correspond to a shift of 1 part in 10^{18}. We can quite safely expect that the effect from nonreciprocal frequency biases is at least one order of magnitude smaller than that. This compares to an operation level, not far from the gravitational wave antennas. The gain medium in the cavity modifies the dispersion effects on the laser frequencies, and that can also influence the nonreciprocal frequency behavior. As a result of this we would expect a laser power dependent bias. Operating the G ring laser at different output power levels reveals an effect of -2.5 μHZ nW^{-1} and this is displayed in figure 6.14. It is an effect that directly results from the differences in the losses between the two senses of wave propagation and shifts the optical frequency systematically, but not at the same rate. The total effect is $\delta f/f = 5 \times 10^{-21}$, clearly a bias value comparable to those of gravitational wave detectors.

Like for every other measurement device, there are classes of systematic measurement errors in the ring laser gyro design. We typically distinguish between three major groups of errors, namely, nullshift, misalignment and lock-in. Let us take a closer look at them. Figure 6.15 illustrates these typical error sources graphically. The first group of errors in this diagram is the scale factor or misalignment error as depicted in figure 6.15(A). This error occurs, when the gyro is operated with a misalignment. That means that the angle θ in equation (6.18) is taken incorrectly. Alternatively, it could

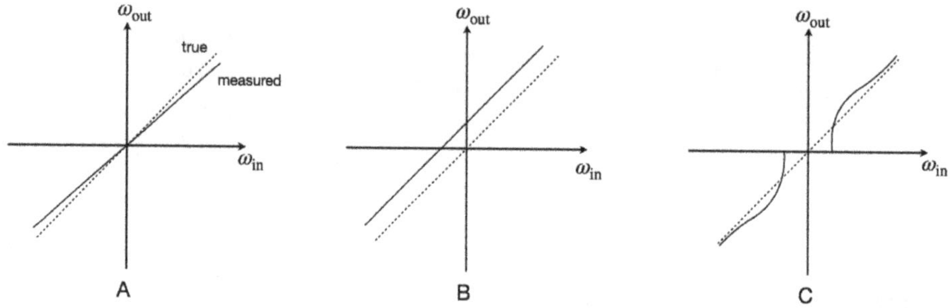

Figure 6.15. Groups of error sources in a ring laser gyroscope. Misalignments and scale factor errors do not transform the experienced rotation rate to the correct measurement value (A). The nullshift error would still show a measurement value even when there is no rotation (B) present. Coupling between the two laser frequencies in the ring cavity causes the signal to disappear, although the instrument is still very slowly turning (C).

also mean that the scale factor has been determined either too large or too small. The dashed line in the diagram indicates the true value. Commercial gyroscopes are calibrated for this kind of errors on a rotary table. This procedure is not possible for large geodetic ring laser gyros. So this error remains as a critical issue in the geodetic ring laser operations. To make sure that we use the correct value for the scale factor, we need to determine the perimeter, the wavelength and the area correctly [9]. Utilizing an optical frequency comb and the determination of the free spectral range of the gyro cavity provides an estimate of the true scale factor that is realistically within 10–100 ppb, depending on how planar the beam path is inside the cavity. To establish the correct latitude is also an elaborate process. Even if the location on the Earth crust has been established with a careful survey, this does not mean that the gyro operates under this angle. The beam path inside the gyroscope may either be skewed or tilted (or in the worst case both of that). The laser beam in a large gyro is of the order of 2–4 mm wide. So, where inside the laser mode is the actual plane of reference located and accessible with good reliability? We have to expect that we cannot obtain θ as exact as we require. This effect can be summarized as the area–orientation ambiguity. When we rewrite equation (6.18) for a system strapped down to Earth and without the nonreciprocal laser frequency contribution, we get

$$\Delta f(t) = \frac{4}{\lambda P} \cdot \frac{2\pi}{86164.1\mathrm{s}} A \sin(\theta(t) + \delta). \tag{6.19}$$

While we can measure λ and P with sufficient accuracy, we have to determine the misalignment error δ and it does not matter whether it is the angle or the area that we can not establish well enough, as long as the value of δ is small. Obtaining δ from observations may prove to be possible and sufficient. This would then provide a calibration for the experimental setting. At this point in time it is an open question if this is sufficient.

The second category of errors are the nullshift errors, which is illustrated in figure 6.15(B). They originate from an offset in the laser frequencies. Normally this affects both laser beams. The nonreciprocal frequency shift that we have discussed above provides a typical case for the nullshift. In the presence of a nullshift offset, one would still read a small value for the measured rotation rate, although the instrument does not experience any rotation. Alternatively and more natural, the extracted rotation rate is slightly larger in value than it actually is. The nullshift is usually related to the nonlinear properties of the laser gain medium in the ring cavity. The nullshift is typically in the range of about 1 ppm for large ring lasers and can be modeled [31]. The final group of errors is more important. It is caused by laser frequency pulling or pushing and is indicated in figure 6.15(C). Two oscillators operating on close frequencies and in the presence of some (unintended) coupling, tend to synchronize in frequency. This effect is present in mechanical systems, in electrical oscillators as well as in lasers. While the coupling in mechanical clocks comes from vibrations, the coupling in a ring laser gyro is caused by light scattered from the turning mirrors inside the cavity. This effect is known as 'backscatter coupling'. A full treatment of this topic is outside the scope of this book but can be found in [10]. It is the largest error in the raw ring laser readings with typical values around 12 ppm. Non of the error sources, discussed above is constant with time. Therefore, we have to constantly track them. For that purpose we operate two photodetectors that constantly measure the intensities of both oscillating laser frequencies in the cavity. This is shown in (7) of figure 6.11(B). Scattering light from one sense of beam propagation into the other causes a beat note with the Sagnac frequency on each laser beam. If there is a lot of light scattered, this beat note has a large amplitude. In the case of weak coupling, the amplitude is small. By monitoring the actual brightness of the beam and the amplitude of the beat note, the appropriate correction can be computed [10]. From the low scatter of the signals shown in figure 6.16 we can see that we can obtain the corrections with good fidelity. The error limit for the G ring laser normalized to the value Ω_E of Earth rotation when all effects are taken together currently is believed to be better than $\Delta\Omega/\Omega_E \leqslant 10^{-7}$, while the resolution is still one order of magnitude higher [23].

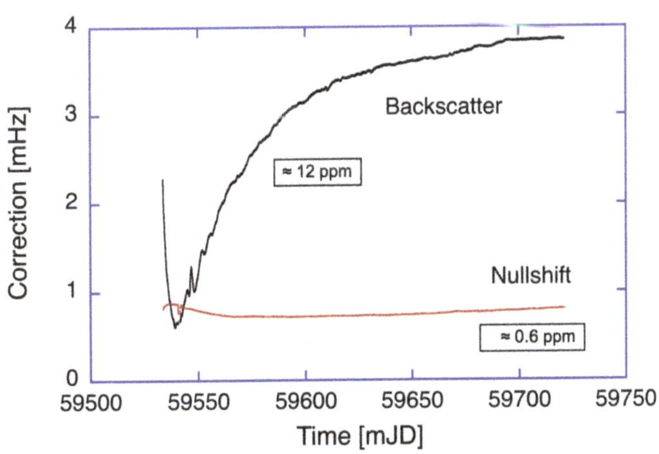

Figure 6.16. The effect of backscatter and nullshift errors in the observations of the large ring laser G as a function of time. While the effect of backscatter coupling causes an offset in the rate of 12 ppm, the nullshift is fairly constant and contibutes with less than 1 ppm.

6.4.3 Ring laser observations

At a first glance we expect the ring laser to observe purely rotations, since we know that such a device is entirely insensitive to translations. However, this turns out to be a hasty conclusion. When we look at the ring laser equation (6.18) more closely, we can identify more contributors to a valid and desired measurement signal. There is the orientation of the laser plane with respect to the Earth rotation axis. Slight tilts of the entire sensor cause the angle θ to change, which in turn varies the observed beat note. The solid Earth tides, induced by the lunar attraction, are causing changes in tilt of the order of 50 nrad and an uplift by about 30 cm in amplitude twice a day. While the former changes the orientation and hence the observed ring laser beat note, the latter is invisible because ring lasers do not feel translations. Figure 6.17 provides an example of solid Earth tides in the North–South direction. This is the direction where the effect is the largest for a horizontally orientated ring laser at midlatitudes. A detailed description of this effect observed by a ring laser can be found in [27]. It is important to note how small the angular change can be and still produce a significant signal contribution on a sensitive gyroscope. Apart from tidal tilt, we can also see effects from variable ocean loading due to oceanic tides. These effects are about one order of magnitude smaller than the solid Earth tides for a large ring laser several hundred km inland from an ocean, such as the G ring laser in Bavaria. The gravitational attraction of Sun and Moon gives rise to polar motion effects. Since the mass distribution of the Earth is inhomogeneous and the momentum axis does not coincide with the figure axis of the Earth, it produces a wobble of the instantaneous rotation axis relative to the body of the Earth. This effect is known as 'forced nutation', 'diurnal polar motion' or 'Oppolzer terms'. Geometrically, the amplitude of the motion of the actual rotation axis is in the range between several cm and 60 cm (figure 6.18).

The result is an effective change in the North South and East–West tilt. A horizontal ring at midlatitudes is mostly sensitive to the North–South tilt, as this

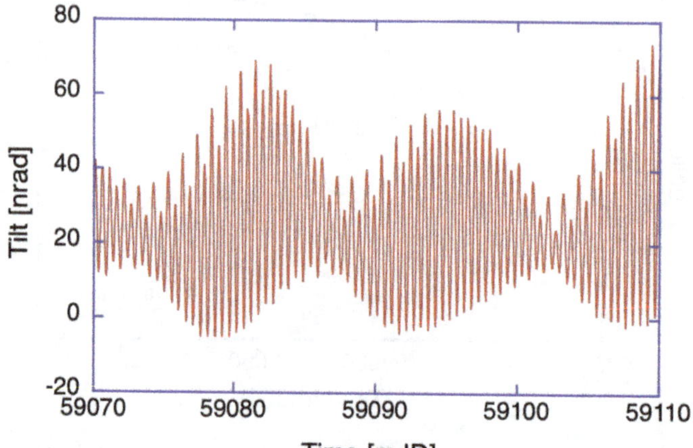

Figure 6.17. Example of North–South tilt angles caused by the deformation induced by the solid Earth tides.

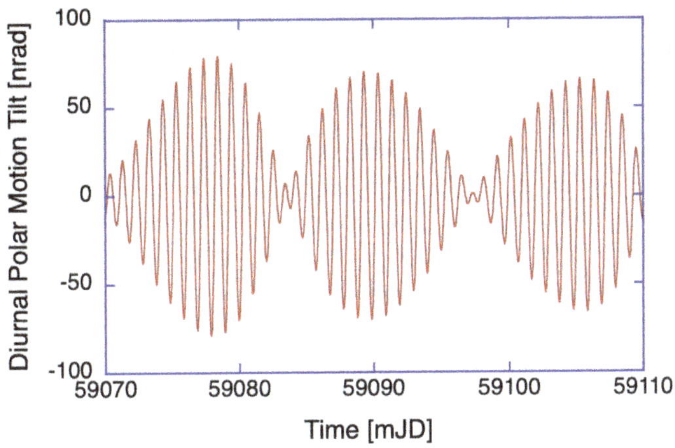

Figure 6.18. Example of the diurnal polar motion caused by forced nutation in the North–South direction.

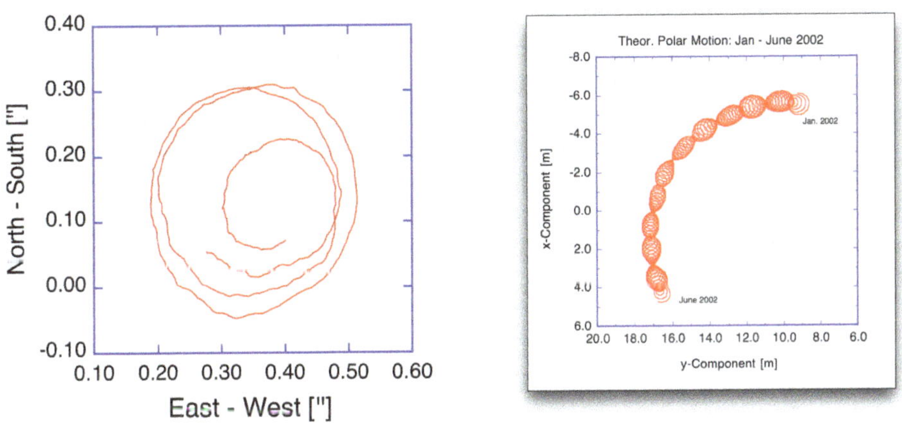

Figure 6.19. The Chandler motion from the C04 series of the IERS between 2022 and 2025 is presented in the units of seconds of arc (left). The diameter of the circles is about 8 m. Superimposed on that is the diurnal polar motion component, here shown for a small segment and in the units of meters (right).

changes the angle θ effectively. This effect is treated in detail in [28] for ring laser observations. The largest effect on the polar motion is caused by the Chandler wobble, which is a free oscillation of the Earth in response to noise excitation of the Earth crust from the wind and the ocean waves.

The Chandler motion shifts the Earth rotation axis around in a roughly circular pattern that has a diameter of approximately 8 m and one circle takes about 434 days to complete. The Chandler motion during the years 2022 until 2025 is shown in figure 6.19. It is taken from the C04 observation series published by the 'International Earth Rotation and Reference System Service' (IERS) and at the time of writing it can be found at [1]. The left side of the figure shows the Chandler motion in the units of seconds of arc. Per convention, the x-axis is aligned with the Greenwich meridian and the y-axis is pointing to 90° West. The right side of the figure exhibits a shorter section of the Chandler motion with the diurnal polar motion superimposed. In this case, the offsets are given in the units of meters for better illustration. The ‘beads-like’ structure is caused by the lunar orbit, and the time offset between consecutive nodes corresponds to half of the lunar orbit, or 14 days [20, 26]. The Chandler motion also exhibits some variation from one circle to the next, which is caused by the slightly eccentric orbit of the Earth around the Sun. Since the annual cycle takes nearly 365.25 days to complete, this periodic change beats with the 434-day period of the Chandler motion, producing a pattern of nodes and antinodes, however, on much longer time scales.

The next contributor to the ring laser observables is a unexpected one and comes at a bit of a surprise. It is the result of the precession and nutation motion of the Earth. The rotation axis of the Earth is tilted by 23.5° against the ecliptic and the gravitational attraction of Sun and Moon on the slightly oblate body of the Earth causes a small force on the rotation vector, trying to pull the rotation vector upright. This causes a precession motion of the rotation vector, corresponding to nearly 26 000 years to complete one cycle. The motion of the Earth rotation axis covers an angle of 50″ over the time of one whole year [6, 15]. One would not expect that to be within the resolution of a large ring laser gyroscope, but it is. The precession and nutation motion are constantly pushing the rotation axis of the Earth further along its considerable cycle and this produces distinct signals in the diurnal frequency band, just as the forced nutation does [23]. Figure 6.20 illustrates these contributions. The precession signal is almost constant in amplitude and biased slightly negative, relative to zero. The nutation acts on much shorter time scales and contains many different contributors, with the 18.6 years of the lunar cycle and the fortnightly signal due to the lunar orbit the largest components. This is clearly visible in the bottom panel of figure 6.20.

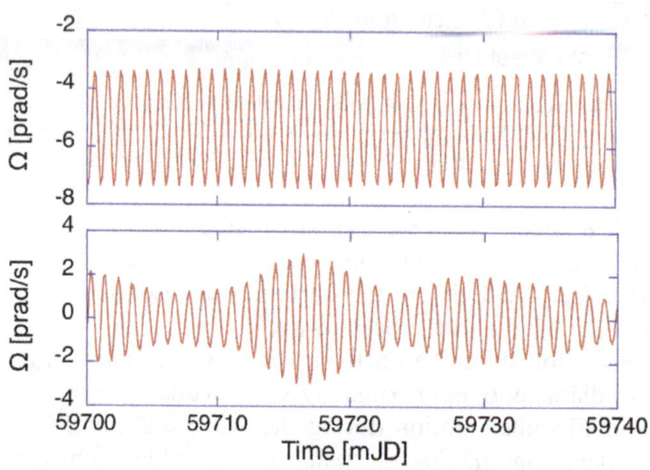

Figure 6.20. The contribution of the precession (top) and nutation (bottom) to the ring laser measurement in the diurnal frequency band. While the precession is constant and negatively biased, the nutation term is composed of many frequency components with the 18.6 years/cycle and the 14 days/cycle signal is the largest in amplitude.

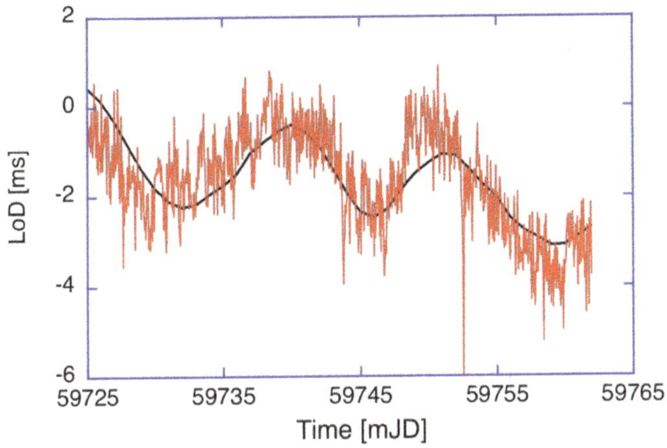

Figure 6.21. Comparison of the Length of Day measurements between high-resolution ring laser measurements and the GNSS- and VLBI-derived observations with one data point per day and some smoothing applied.

The Earth constantly exchanges rotational energy between the solid part of the Earth and the liquids, namely the hydrosphere, the atmosphere and at longer time scales the cryosphere as well. As a consequence the Earth is slightly speeding up at times and then slows down again. The measurements of these effects provides boundary conditions for models that are computing these very small effects of about +2 ms in 14 days or expressed differently at the level of 1 part in 10^8 with respect to the rotation rate of the Earth [7, 24]. Most dominant are seasonal changes, which occur over several months. A prominent example is the El Nino Southern Oscillation, which usually lasts for more than half a year. The current stability of the large ring laser G is sufficient to identify the short-term Length of Day (LoD) signal with a period of 14 days. Figure 6.21 depicts a comparison of the measurements taken by ring laser with a temporal resolution of 1 h and the IERS provided measurement results, which depend on GNSS observations, stabilized by VLBI measurements. There is one data point per day and some smoothing applied to the IERS data.

In this section we have seen that the Earth rotation rate, which an inertial sensor records, is composed of many different components, acting at very different time scales. Interestingly, most of the signals become visible as tilt effects rather than actual variations in the rotation rate of the Earth. This means that a stable and sensitive measurement of the tilts acting on the gyroscope is a mandatory part of the operation. For a single horizontal ring laser, this requires a sensitive and long-term stable tiltmeter. However, a three-component stable ring laser would be self-contained. The tilt on one axis corresponds to a rotation on another axis and therefore it would be possible to recover this motion as long as the beam path is not affected.

6.5 Ring laser in geodesy

The transformation between the celestial and the terrestrial reference frame requires precise knowledge of the rotational velocity of the Earth as well as the orientation of the rotation axis in space. Ideally, it would even require a continuous observation. There are essentially two ways in which the instantaneous earth orientation parameters can be observed. One can measure angles as a function of time between fixed objects in space. This could be regarded as a star compass. Alternatively, one could use a highly sensitive gyroscope rigidly strapped down to Earth to perform this task. This would then be an inertial realtime compass. Applications in the modern society, like autonomous harvesting, precision navigation and ultimately autonomous traffic control and driving, require that any two points on Earth can be related to each other with an uncertainty of one centimeter on a global scale and in realtime, which calls for a measurement resolution of 1 part in 10^9 and that is a formidable demand. For the star compass this is equivalent to an angular resolution in the range of 10–100 μas. For the estimation of the LoD signal a resolution of 50–100 μs is required. A network of much more than 1000 permanent GNSS stations coordinated by the International GNSS service (IGS) performs this task on a daily basis. In order to ensure the stability, recurrent observations by the VLBI network are required. These measurements provide the link to the ICRF, established by a catalog of defining quasar positions. All of this marks a very involved process and lacks a consistency check by an independent measurement technique. Large monolithic ring lasers have now reached a point of maturity, where they are useful for the application in space geodesy. So they could fill this gap. Nevertheless, the requirement of a resolution in the ppb regime and a long-term stability of years if not decades is a stringent demand. While the necessary resolution can be obtained with the current ring laser technology, the required long-term stability is not available from an optical cavity alone. At the time of writing, only the monolithic single-component G ring laser at the Geodetic Observatory Wettzell has demonstrated the necessary measurement resolution. For the full estimate of all the Earth orientation parameters a triad of G-type ring lasers would be required and this alone is not enough. This triad would need to be rigidly attached to the Earth's crust in a way that the local measurement of the Earth's orientation parameters would be correctly representing these global quantities. This cannot be expected. VLBI is an interferometric technique, just as a ring laser (Sagnac interferometer). It operates on signals with much larger wavelength, namely about six orders of magnitude. However, the distance over which the interferogram is taken by the VLBI antennas, can be as large as 10 000 km. While this makes the evaluation of the radio interferogram more difficult, it also relaxes the requirements for the sensor stability. Furthermore, the VLBI measurements are made with sufficient redundancy to identify inconsistencies in the measurements. However, these measurements are obtained from discrete sessions and they are not continuous.

A network of similar large ring laser gyroscopes would provide the necessary redundancy for inertial rotation sensing and they would operate autonomously and continuously. Despite of this, we cannot expect such a network to provide a consistent and stable reference to the Earth crust in a global sense. Furthermore, a ring laser, like the G ring laser weighs several tons and extends over several meters in all directions. Eventually, it is the plane of the centers of the laser beams, against which the measurements are taken and this does not necessarily align with the

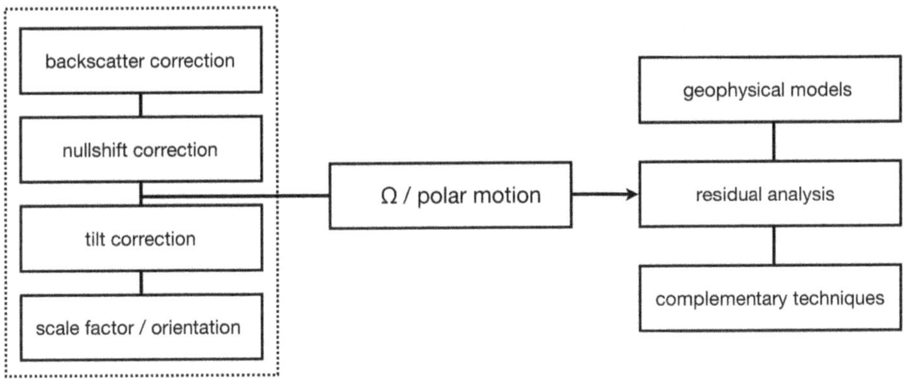

Figure 6.22. The dotted box on the left includes the sensor model which defines the scale factor and the initial orientation of the gyroscope. It also provides corrections for systematic sensor errors and maintains the orientation. The result is a time series of earth rotation and polar motion (center). On the right side we find the geophysical models and the observations from the other techniques of space geodesy, which can now be compared.

enclosing monolithic body of the gyroscope component. So we cannot just simply take a gyroscope and assume that we measure valid earth orientation parameters straight out of the box. The instrumentation will be more complex and we need to operate several auxiliary sensors along with the Sagnac signal. The scheme outlined in figure 6.22 is currently the best available approach.

A number of auxiliary measurements on the ring laser hardware, such as the instantaneous orientation against the local g, the optical frequency comb measurements for the correct scale factor [9], the estimation of the backscatter coupling [10] and the nullshift offset [31] are required to obtain the correct instantaneous rotation rate and the signature of the polar motion. Once corrected, the measurements can now be compared to the available geophysical models and to the observations of the other space geodetic techniques. We have performed this analysis scheme on the G ring laser measurements. In order to use the correct orientation, we have established the scale factor and then used actual measurements to infer the instantaneous orientation of the G ring laser. From the Bernese GNSS software [5] we obtained the transformation from the celestial to the terrestrial reference frame and from the IERS we used the C04 time series [1]. When we subtracted the modeled rotation rate and the Earth orientation parameters from the ring laser measurements, most of the signals disappeared, thus indicating a consistent measurement. Computing a spectrum from the residuals, provided the diagram depicted in figure 6.23. We see some residual signals in the diurnal and semidiurnal frequency band. This comes from a discrepancy between the models and the measurements.

The strongest signal in the diurnal band features a value of less than 4 ppb, which suggests that the evaluation approach already provides excellent data quality. The discrepancy can be due to remaining systematic errors in the ring laser measurements or could be due to some small model deficits or small inconsistencies in the C04 time series. Most likely, these residuals have more than one distinct cause. We

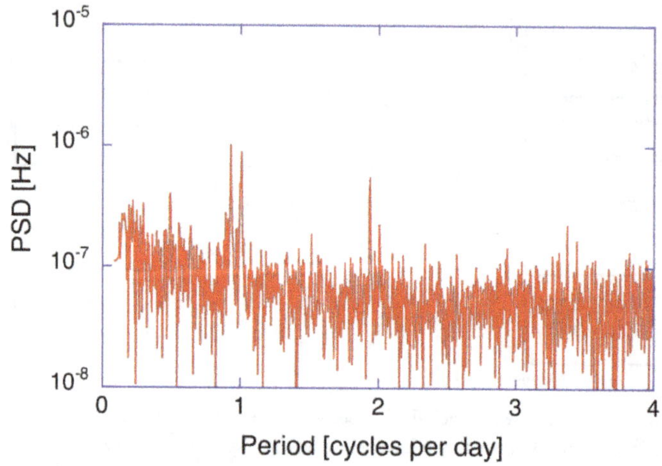

Figure 6.23. The calcualted spectrum taken from the residuals of the ring laser measurements, once the contributions from the geophysical models were subtracted from the corrected ring laser measurements.

have looked at some alternatives and the general experience is that there are small improvements, say in the diurnal band, and at the same time the power in the semidiurnal band increases, or vice versa. The current quality of the ring laser observations are putting us into a comfortable position, since they provide an entirely independent and highly sensitive set of observations at the resolution of about 1 ppb.

While the most consistent and best resolved ring laser measurements have been taken with a monolithic gyro with a size of 4 × 4 m, we have also developed the large four-component ROMY ring laser array, composed of four triangular laser cavities, where each side is 12 m long. This ring laser array features the shape of a tetrahedron with the tip in the ground [11]. It is arranged within an underground concrete monument. The laser beams are propagating inside stainless steel tubes. The effective scale factor of each gyro in the array is about twice that of G. At the time of writing there are no controls that can stabilize the resonator geometry sufficiently. This leads to some scale factor instabilities and beam walk effects. Although it was possible to reconstruct the full earth rotation vector from the measurements [8], the instability is still too large to make it useful for space geodesy. However, this approach is probably the best viable path into the future. The G ring laser has demonstrated that Sagnac interferometry provides the necessary resolution and can reach adequate stability, but a full three-component system would be prohibitively expensive. ROMY does not yet have the required stability, but the necessary path forward is clear and a matter of concentrated effort and funding.

References

[1] Earth orientation data 2025 https://datacenter.iers.org/data/latestVersion/EOP_20_C04_one_file_1962-now.txt (online; accessed 3 July 2025)

[2] Aronowitz F 1971 The laser gyro *Laser Appl.* **1** 133–200

[3] Bilger H, Stedman G, Li Z, Schreiber K U and Schneider M 1995 Ring lasers for geodesy *IEEE Trans. Instrum. Meas.* **44** 468–70

[4] Chao B F 1991 As the world turns, II *EOS Trans. Am. Geophys. Union* **72** 550–1

[5] Dach R, Lutz S, Walser P and Fridez P 2015 *Bernese GNSS Software version 5.2 User manual* (Bern Open Publishing)

[6] Ekman M 1993 A concise history of the theories of tides, precession-nutation and polar motion (from antiquity to 1950) *Surv. Geophys.* **14** 585–617

[7] Freedman A P, Steppe J A, Dickey J O, Eubanks T M and Sung L Y 1994 The short-term prediction of universal time and length of day using atmospheric angular momentum *J. Geophys. Res.* **99** 6981–96

[8] Gebauer A *et al* 2020 Reconstruction of the instantaneous Earth rotation vector with sub-arcsecond resolution using a large scale ring laser array *Phys. Rev. Lett.* **125** 033605

[9] Hurst R B, Mayerbacher M, Gebauer A, Schreiber K U and Wells J -P R 2017 High-accuracy absolute rotation rate measurements with a large ring laser gyro: establishing the scale factor *Appl. Opt.* **56** 1124–30

[10] Hurst R B, Rabeendran N, Schreiber K U and Wells J -P R 2014 Correction of backscatter-induced systematic errors in ring laser gyroscopes *Appl. Opt* **53** 7610–8

[11] Igel H *et al* 2021 ROMY: a multicomponent ring laser for geodesy and geophysics *Geophys. J. Int.* **225** 684–98

[12] Lodge O 1897 Experiments on the absence of mechanical connexion between ether *Phil. Trans. R. Soc.* **189** 149–66

[13] Macek W M and Davis D T M 1963 Rotation rate sensing with traveling-wave ring lasers *Appl. Phys. Lett.* **2** 67–8

[14] Maiman T H 1960 Stimulated optical radiation in ruby *Nature* **187** 493–4

[15] Mathews P M, Herring T A and Buffett B A 2002 Modeling of nutation and precession: New nutation series for nonrigid Earth and insights into the Earth's interior *J. Geophys. Res.* **107** 3-1–26

[16] Meschede D 2017 *Optics, Light and Lasers* (Wiley)

[17] Michelson A A 1925 The effect of the Earth's rotation on the velocity of light, part I *Astrophys. J.* **61** 137

[18] Michelson A A and Gale H G 1925 The effect of the Earth's rotation on the velocity of light, part II *Astrophys. J.* **61** 2

[19] Michelson A A and Morley E W 1881 *On the Relative Motion of the Earth and the Luminiferous Ether* (SPIE Milestone Series, vol 28) pp 450–8

[20] Na S -H, Roh K -M, Cho J, Yoo S -M, Choi B and Yoon H 2019 Chandler wobble and free core nutation: theory and features *J. Astron. Space Sci.* **36** 11–20

[21] Rodloff R 1990 Concept for a high precision experimental laser gyroscope system 'ELSy' (laser gyro goniometry) *Physica* C **12** 67–74

[22] Sagnac G 1913 Laether lumineux demontre par laeffet du vent relatif daether dans un interferometre en rotation uniforme *C. R. Physique* **157** 708–10

[23] Schreiber K U, Hugentobler U, Kodet J, Stellmer S, Klügel T and Wells J-P R 2025 Gyroscope measurements of the precession and nutation of Earth's axis *Sci. Adv.* **11** eadx6634

[24] Schreiber K U, Kodet J, Hugentobler U, Klügel T and Wells J-P R 2023 Variations in the Earth as rotation rate measured with a ring laser interferometer *Nat. Photon.* **17** 1054–8

[25] Schreiber K U and Wells J -P R 2023 *Rotation Sensing with Large Ring Lasers: Applications in Geophysics and Geodesy* (Cambridge University Press)

[26] Schreiber K U, Klügel T, Wells J -P R, Hurst R B and Gebauer A 2011 How to detect the chandler and the annual wobble of the Earth with a large ring laser gyroscope *Phys. Rev. Lett.* **107** 173904

[27] Schreiber K U, Klügel T and Stedman G 2003 Earth tide and tilt detection by a ring laser gyroscope *J. Geophys. Res.* **108** 2132

[28] Schreiber K U, Velikoseltsev A, Rothacher M, Klügel T, Stedman G and Wiltshire D L 2004 Direct measurement of diurnal polar motion by ring laser gyroscopes *J. Geophys. Res.* **109** 6405

[29] Schreiber K U and Wells J -P R 2013 Invited Review Article: Large ring lasers for rotation sensing *Rev. Sci. Instrum.* **84** 041101

[30] Stedman G E 1997 Ring-laser tests of fundamental physics and geophysics *Rep. Prog. Phys.* **60** 615

[31] Virgilio A D V D, Beverini N, Carelli G, Ciampini D, Fuso F and Maccioni E 2019 Analysis of ring laser gyroscopes including laser dynamics *Eur. Phys. J D* **79** 041101

[32] Wilkinson J R 1987 Ring lasers *Prog. Quantum Electron.* **11** 1–103

Index

doi:10.1088/978-0-7503-2032-0ch7

www.ingramcontent.com/pod-product-compliance
Lightning Source LLC
LaVergne TN
LVHW062239200926
848484LV00006B/1574